Foundations of geometry

Selected proceedings
of a conference

Foundations of geometry

Selected proceedings of a conference

Edited by P. Scherk

University of Toronto Press
Toronto and Buffalo

LIBRARY OF CONGRESS CATALOGING IN PUBLICATION DATA

Main entry under title:

Foundations of geometry

Conference held at the University of Toronto,
July 17-Aug. 18, 1974.
1. Geometry - Foundations - Congresses. I. Scherk,
Peter. II. Toronto. University.
QA681.F74 516 75-42127
ISBN 0-8020-2216-2

CONTENTS

PREFACE

In 1974 the University of Toronto sponsored a unique experiment. For more than four weeks, from July 17 to August 18, a group of outstanding workers in the foundations of geometry lived together on the campus of the university, gave and attended lectures, and met informally.

Three longer series of sixteen lectures each were given by F. Bachmann (Kiel University, West Germany) on Hjelmslev groups, by R. Lingenberg (Karlsruhe, West Germany) on S-groups, and by J. Tits (Collège de France, Paris) on generalized polygons. Two of these are being prepared for publication as separate volumes.

In addition, fifteen mathematicians gave not the usual brief accounts of their work but detailed reports which required up to seven hours of lecturing. Thirteen of these reports appear in this volume. The other two reports, by E. Sperner (Hamburg University, West Germany) on affine structures and by J. Tits on quadratic forms and the geometry on quadrics, unfortunately could not be included.

The editor wishes to thank the University of Toronto Press and the Blythe Foundation for making publication of the volume possible. The papers represent an invaluable basis for further progress in the field, as has already been demonstrated by the appearance of one important paper which received its stimulus from the conference.

CONTRIBUTORS

BARLOTTI, ADRIANO
Università di Bologna, Istituto Matematico, Bologna, Italy

COFMAN, JUDITA
Fachbereich Mathematik der Johannes Gutenberg Universität, 6500 Mainz, Postfach 3980, West Germany

GÖTZKY, MARTIN
Mathematisches Seminar der Christian Albrechts Universität, 2300 Kiel, Neue Universität, Angerbau A1, West Germany

HERING, CHRISTOPH
Mathematisches Institut der Universität Tübingen, 7400 Tübingen 1, Auf der Morgenstelle 10, West Germany

KARZEL, HELMUT
Institut für Geometrie der Technischen Universität, 8000 Munich 2, Arcisstrasse 21, West Germany

LORIMER, J.W.
Department of Mathematics, University of Toronto, Toronto, Ontario, M5S 1A1, Canada

LÜNEBURG, HEINZ
Fachbereich Mathematik der Universität Kaiserslautern, 675 Kaiserslautern, Pfaffenbergstrasse, West Germany

MENDELSOHN, ERIC
Department of Mathematics, University of Toronto, Toronto, Ontario, M5S 1A1

OSTROM, T.G.
Department of Mathematics, Washington State University, Pullman, Wash. 99163, U.S.A.

SALOW, EDZARD
28 Bremen, Katrepelerstrasse 7, West Germany

SEIDEL, J.J.
Department of Mathematics, Technological University Eindhove
PO Box 513, Eindhoven, The Netherlands

STRAMBACH, KARL
Mathematisches Institut der Universität Erlangen-Nürnberg,
852 Erlangen, Bismarckstrasse 1½, West Germany

YAQUB, JILL C.D.S.
Department of Mathematics, Ohio State University, 231 West
18th Avenue, Columbus, Ohio 43210, USA

FOUNDATIONS OF GEOMETRY

Selected proceedings of a conference

COMBINATORICS OF FINITE PLANES AND OTHER FINITE GEOMETRIC STRUCTURES

A. Barlotti

In the following pages, we survey recent results on some of the topics considered in section 3.2 of P. Dembowski's book *Finite Geometries* (1968; for other references on this subject, see A. Barlotti, 1974).

1. SYSTEMS AXIOMATIZING FINITE PLANES AND OTHER FINITE STRUCTURES

Many authors have contributed to the study of systems of axioms for finite projective and affine planes (see Dembowski, 1968, pp. 138-139). Using the notation of Dembowski, we state the following results.

THEOREM A (A. Basile, 1970). *Let* $n > 1$ *be an integer. Then the following is a complete list of minimal axiom systems, each one of which will ensure that a non-degenerate incidence structure is a finite projective plane of order* n:

(a) (1'),(2"),(4"),(5'), (a') (1"),(2'),(3"),(6');

(b) (1'),(2"),(4"),(6'), (b') (1"),(2'),(3"),(5');

(c) (1'),(2"),(3'),(6"), (c') (1"),(2'),(4'),(5");

(d) (1'),(2"),(5'),(6"), (d') (1"),(2'),(5"),(6');

(e) (1'),(2"),(6), (e') (1"),(2'),(5);

(f) (2'),(3"),(4"),(5'), (f') (1'),(3"),(4"),(6');

(g) (1'),(3"),(4"),(5'), (g') (2'),(3"),(4"),(6');

(h) (1"),(2'),(3"),(4'),(6"),

(h') (1'),(2"),(3'),(4"),(5");

(i) (2"),(3),(4'),(6"), (i') (1"),(3'),(4),(5");

(j) (1"),(3),(4'),(6"), (j') (2"),(3'),(4),(5");

(k) (2'),(3"),(5'),(6"), (k') (1'),(4"),(5"),(6');

(ℓ) (1'),(3"),(5'),(6"), (ℓ') (2'),(4"),(5"),(6');

(m) (3"),(4'),(5'),(6"), (m') (3'),(4"),(5"),(6');

(n) (2'),(3"),(6) (n') (1'),(4"),(5);

(o) (1'),(3"),(6), (o') (2'),(4"),(5);

(p) (3"),(4'),(6), (p') (3'),(4"),(5);

(q) (1"),(3'),(4'),(5"),(6"),

(q') (2"),(3'),(4'),(5"),(6");

(r) (2'),(5),(6"), (r') (1'),(5"),(6);

(s) (1'),(5),(6"), (s') (2'),(5"),(6);

(t) (4'),(5),(6"), (t') (3'),(5"),(6);

(u) (3'),(5),(6"), (u') (4'),(5"),(6).

THEOREM B (P. Brutti, 1969; G. Pabst, 1974). *Let* $n > 2$ *be an integer. Then the following is a complete list of minimal axiom systems, each one of which will ensure that* a non-degenerate incidence structure is a finite affine plane of order n:

(a) (3),(5),(9);

(b') (5),(7),(9");

(c') (5),(8),(9);

(d') (3'),(5"),(7"),(9);

(e) (3"),(5'),(7'),(9");

(f) (3'),(5"),(8"),(9);

(g') (3"),(5'),(8'),(9");

(k) (5"),(7"),(8'),(9');

(ℓ) (5'),(7'),(8"),(9");

(m) (3'),(5"),(7'),(8"),(9");

(n) (3"),(5'),(7"),(8').

Clearly additional hypotheses will modify the above systems. C. Bernasconi (to appear, a) studied the problem with the additional assumption that the structure is connected, and M. Crismale (to appear) changed (3'), (3"), (4'), (4") to axioms which require that in (3) and (4) "=" be replaced by "≤" or "≥" in average (with the symbols in Dembowski [1968], $v_1 \leq n + 1$, $v_1 \geq n + 1$, $b_1 \leq n + 1$, $b_1 \geq n + 1$).

Similar systems of axioms have been studied for other finite structures: for inversive planes (R. Bumcrot, private communication) and for projective designs (C. Bernasconi, to appear, b).

In connection with the above questions, see P. Biscarini (to appear), N.G. de Bruijn and P. Erdös (1968), R. Magari (1964), and I. Reiman (1963, 1968). In particular we wish to point out that D.A. Drake (1974) has proved that in a finite Hjelmslev plane the parallel axioms cannot be replaced by cardinality assumptions.

2. REPRESENTATION OF NETS AND PLANES

We can use the representation of nets and planes by sets of mutually orthogonal latin squares to study the propertie of the plane (see R.C. Bose and K.R. Nair, 1941; G. Pickert 1955).

For recent results in this field, see A. Barlotti (to appear), J. Dénes and A.D. Keedwell (1974), and P. Hohler (1970, 1972).

3. GALOIS GEOMETRIES

Galois geometries are the study of the sets of points in finite spaces. In the following, PG(r, q) will denote an r-dimensional projective space of order q. If $r = 2$, the symbol PG(2, q) will be used only for a desarguesian plane whereas $\pi(q)$ will denote any projective plane of order q.

A (k; n)-arc of $\pi(q)$ is a set of k points of $\pi(q)$ such that n is the largest number of them which are collinear. In a given plane, a (k; n)-arc is complete if there does not exist a (k'; n)-arc which contains it with

$k' > k$.

(k; 2)-arcs are simply called k-arcs; i.e. a k-arc of $\pi(q)$ is a set of k points of $\pi(q)$ such that no three of them are collinear.

3.1. Some properties of k-arcs

In the last twenty years, k-arcs have been studied extensively, especially in PG(2, q). We shall recall here some of their properties.

A line of $\pi(q)$ is called a secant, a tangent, or an external line to a k-arc K if it meets K in two, one, or zero points. It is very easy to prove that in $\pi(q)$ the number of tangents of a k-arc passing through a point not on the arc has the same parity as k.

A question which in general is still open is determining the maximum value of k for which, in a given $\pi(q)$, a k-arc will exist (the packing problem for k-arcs). In the general case, we have the following result:

THEOREM 3.1.1 (R.C. Bose, 1947). *If there exists a* k-*arc in* $\pi(q)$, *then* $k \le q + 1$ *when* q *is odd and* $k \le q + 2$ *when* q *is even*.

The following theorems give the existence of maximal arcs in the desarguesian planes.

THEOREM 3.1.2. *In* PG(2, q), *an irreducible conic is a* (q + 1)-*arc*.

THEOREM 3.1.3 (B. Quist, 1952). *In* $\pi(q)$, *all the tangents of a* $(q + 1)$-*arc are concurrent when* q *is even*.

This implies that in $\pi(q)$, if q is even, then every $(q + 1)$-arc is incomplete and can be uniquely completed to form a $(q + 2)$-arc. The point at which all the tangents of a $(q + 1)$-arc meet is called the nucleus of that $(q + 1)$-arc.

The question of the existence of $(q + 1)$-arcs in every projective plane $\pi(q)$ is still open.

THEOREM 3.1.4 (G. Korchmáros, to appear). *Each Hall plane of odd order contains a* $(q + 1)$-*arc*.

A graphic characterization for conics in PG(2, q), q odd, is given by the following theorem.

THEOREM 3.1.5 (B. Segre, 1955). *In* PG(2, q), *if* q *is odd*, *then every* $(q + 1)$-*arc is an irreducible conic*.

In PG(2, 2^h), we can obtain a $(q + 2)$-arc by adjoining the nucleus to a conic (see Theorem 3.1.3). By deleting a point other than the nucleus from this $(q + 2)$-arc, we obtain a $(q + 1)$-arc which is not a conic provided $q > 5$. For if this arc is a conic, then by the theorem of Bezout, it cannot have more than four points in common with the original conic. The question arises as to whether or not there exist $(q + 2)$-arcs which do not contain a conic. The following theorem provides the answer.

THEOREM 3.1.6 (B. Segre, 1957). In $PG(2, 2^h)$, there exist $(q + 2)$-arcs which do not contain $q + 1$ points forming a conic when $h = 4$ or $h = 5$ or $h \geq 7$.

It is not known if the same property holds for $h = 6$. When $h = 1, 2$, or 3, every $(q + 2)$-arc can be obtained by adding the nucleus to the $q + 1$ points of a conic.

We state now a theorem which shows how a geometrical property of an arc can determine the nature of the plane in which the arc can be embedded:

THEOREM 3.1.7 (F. Buekenhout, 1966). If in $\pi(q)$ there is a $(q + 1)$-arc K such that every hexagon whose vertices are points of K is pascalian, then $\pi(q)$ is pappian and K is a conic.

We wish to point out that the introduction of "segments" in the geometry over a field by B. Segre (1973) leads to results of deep interest in the study of conics.

For more details on the study of k-arcs, the reader is referred to A. Barlotti (1965) and B. Segre (1961, 1965, 1967).

3.2. Some properties of $(k; n)$-arcs

The study of $(k; n)$-arcs, for $n > 2$, is not as developed as the study of k-arcs. The "packing problem" has not been solved, not even for the case of PG(2, q). A. Basile and P. Brutti (1971, 1973) and H.-R. Halder (to appear) have obtained interesting results on this problem with the aim of proving the Lunelli-Sce conjecture (see L. Lunelli and M. Sce, 1964).

In the following, we present a brief list of topics with which (k; n)-arcs are connected or in which they are of interest.

(a) The use of (k; n)-arcs in the construction of geometric structures. R.C. Bose (1958) used a special class of $(p^{3m} + 1, p^{m} + 1)$-arcs to derive BIB-designs. T.G. Ostrom (1962) used arcs to construct Bolyai-Lobachevsky planes and J.A. Thas (1973, 1974) used (k; n)-arcs to construct partial geometries. The connection between arcs and certain (finite) Laguerre-m-structures is also of interest (see H.-R. Halder and W. Heise, to appear).

Structures in high-dimensional finite spaces, corresponding to the k-arcs of the projective plane over the total matrix algebra of the $n \times n$ matrices with elements in GF(q), were used by J.A. Thas (1973) to construct special classes of 4-gonal configurations. See also the constructions due to J. Tits in P. Dembowski (1968).

(b) The relationship between (k; n)-arcs and other subsets of points of $\pi(q)$. A. Bruen and J.C. Fisher (1973) found an interesting connection between complete k-arcs and "blocking sets." (A blocking set S in $\pi(q)$ is a subset of the points of $\pi(q)$ such that every line of $\pi(q)$ contains at least one point in S and at least one point not in S [see, for example, A. Bruen, 1971].)

(c) To obtain information on planes, whose existence is not yet known, by studying (k; n)-arcs in the planes (with the hypothesis that the planes exist). As examples: R.H.F. Denniston (1969) proved that no 6-arc is complete in $\pi(10)$ and A. Barlotti (1971) obtained some (k; n)-arcs in the plane $\pi(12)$ with a subplane of order 3.

3.3. k-caps and the "packing problem"

A set of k distinct points of PG(r, q) ($r \geq 3$), no three of which are collinear, is called a k-cap. We shall denote by m(r, q) the maximum number of points in PG(r, q) that belong to a k-cap. The value of m(r, q) is very important for some applications in coding theory and in statistics. However, it is a very difficult problem to determine m(r, q) ("packing problem") in the general case.

The following are the known values of m(r, q) ($r \geq 3$):

$m(3, q) = q^2 + 1$,

$m(r, 2) = 2^r$,

$m(4, 3) = 20$ (G. Pellegrino, 1970),

$m(5, 3) = 56$ (R. Hill, to appear).

A list of the known bounds for m(r, q) is given in a paper by B. Segre (1967, p.166).

For a wide illustration of further problems on caps, see the papers due to appear by G. Tallini and M. Tallini Scafati.

REFERENCES

Barlotti, A. 1965. Some topics in finite geometrical structures. Lecture Notes, Chapel Hill, N.C.

—— 1971. Alcuni procedimenti per la costruzione di piani grafici non desarguesiani. Conferenze Sem. Mat. Bari 127: 1-17.

—— 1973. Some classical and modern topics in finite geometrical structures. In: A Survey of Combinatorial Theory, edited by J.N. Srivastava et al., pp. 1-14. North Holland Co.

—— 1974. Combinatorics of finite geometries. In: Combinatorics, edited by M. Hall Jr. and J.H. van Lint, Part I, pp. 55-59. Mathematical Centre Tracts 55, Math. Centrum, Amsterdam.

—— (to appear) Alcune questioni combinatorie nello studio delle strutture geometriche. In: Atti Convegno Teorie Combinatorie, Acc. Lincei, Rome 1973.

Basile, A. 1970. Sugli insiemi di proprietà che definiscono un piano grafico finito. Le Matematiche 25: 84-95.

Basile, A. and P. Brutti. 1971. Alcuni risultati sui $\{q(n-1)+1; n\}$-archi di un piano proiettivo finito. Ren. Sem. Mat. Univ. Padova 46: 107-125.

—— 1973. On the completeness of regular $\{q(n-1)+m; n\}$-arcs in a finite projective plane. Geom. Dedicata 1: 340-343.

Bernasconi, C. (to appear) (a) Strutture di incidenza connesse e definizione assiomatica di piani grafici e affini. In: Ann. Univ. di Ferrara.

—— (to appear) (b) Sistemi di assiomi che caratterizzano i disegni proiettivi.

Biscarini, P. (to appear) Sets of axioms for finite inversive planes.

Bose, R.C. 1947. Mathematical theory of the symmetrical factorial design. Sankhya 8: 107-166.

—— 1958. On the application of finite projective geometry for deriving a certain series of balanced Kirkman arrangements. The Golden Jubilee Commemoration Volume, Calcutta Math. Soc.

—— 1973. On a representation of Hughes planes. In: Proc. Internat. Conf. on Projective Planes, edited by M.J. Kallaher and T.G. Ostrom, pp. 27-57. Wash. State Univ. Press.

Bose, R.C. and K.R. Nair. 1941. On complete sets of Latin squares. Sankhya 5: 361-382.
Bruen, A. 1971. Blocking sets in finite projective planes. SIAM J. Appl. Math. 21: 380-392.
Bruen, A. and J.C. Fisher. 1972. Arcs and ovals in derivable planes. Math. Z. 125: 122-128.
—— 1973. Blocking sets, k-arcs and nets of order ten. Advances in Math. 10: 317-320.
Brutti, P. 1969. Sistemi di assiomi che definiscono un piano affine di ordine n. Ann. dell'Univ. Ferrara, ser. VII, 14: 109-118.
Buekenhout, F. 1966. Plans projectifs à ovoides pascaliens. Arch. Math. 17: 89-93.
Crismale, M. (to appear) Sui sistemi minimi di assiomi atti a definire un piano proiettivo finito.
de Bruijn, N.G. and Erdös. 1968. On a combinatorial problem. Indag. Math. 10: 421-423.
Dembowski, P. 1968. Finite Geometries. Springer-Verlag, Berlin-Heidelberg-New York.
Dénes, J. and A.D. Keedwell. 1974. Latin Squares and Their Applications. Academic Press, New York and London.
Denniston, R.H.F. 1969. Nonexistence of a certain projective plane. J. Austral. Math. Soc. 10: 214-218.
Drake, D.A. 1974. Near affine Hjelmslev planes. J. Comb. Theory 16: 34-50.
Halder, H.-R. (to appear) Über symmetrische (k; n)-Kurven in endlichen Ebenen.
Halder, H.-R. and W. Heise. (to appear) On the existence of finite Laguerre-m-structures and k-arcs in finite projective spaces.
Hall, M. Jr. 1959. The Theory of Groups. Macmillan, New York.
Hill, R. (to appear) On the largest size of cap in $S_{5,3}$. Rend. Acc. Naz. Lincei.
—— (to appear) Caps and groups.
Hohler, P. 1970. Verallgemeinerung von orthogonalen lateinischen Quadraten auf höhere Dimensionen. Diss. 4522, Eidg. Technischen Hochschule Zürich.
—— 1972. Eigenschaften von vollständigen Systemen orthogonaler Lateinischer Quadrate, die bestimmte affine Ebenen repräsentieren. J. Geom. 2: 161-174.
Korchmáros, G. (to appear) Ovali nei piani di Hall di ordine dispari.
Lunelli, L. and M. Sce. 1964. Considerazioni aritmetiche e risultati sperimentali sui {k; n}-archi. Rend. Ist. Lombardo (A) 98: 3-52.
Magari, R. 1964. Sui sistemi di assiomi "minimali" per una data teoria. Boll. UMI 19: 423-435.
Menichetti, G. (to appear) q-archi completi nei piani di Hall di ordine $q = 2^k$.

Ostrom, T.G. 1962. Ovals and finite Bolyai-Lobachevsky planes. Amer. Math. Monthly 69: 899-901.
Pabst, G. 1974. Private communication. (The result is in the author's master thesis.) See also "Quantitative axioms for affine planes" in Abstract of Communications ICM, Vancouver, 1974, p. 109.
Pellegrino, G. 1970. Sul massimo ordine delle calotte in $S_{4,q}$. Le Matematiche 25: 149-157.
—— 1972. Procedimenti geometrici per la costruzione di alcune classi di calotte complete in $S_{r,3}$. Boll. UMI (4) 5: 109-115.
Pickert, G. 1955. Projektive Ebenen. Springer-Verlag, Berlin-Göttingen-Heidelberg.
Quist, B. 1952. Some remarks concerning curves of the second degree in a finite plane. Ann. Acad. Sci. Fenn. 134: 1-27.
Reiman, I. 1963. Su una proprietà dei piani grafici finiti. Rend. Acc. Naz. Lincei 35: 279-281.
—— 1968. Su una proprietà dei due disegni. Rend. Mat. e Appl. 76-81.
Segre, B. 1955. Ovals in a finite projective plane. Canad. J. Math. 7: 414-416.
—— 1957. Sui k-archi nei piani finiti di caratteristica due. Rev. Math. Pures Appl. 2: 289-300
—— 1961. Lectures on Modern Geometry. Cremonese, Rome.
—— 1965. Istituzioni di geometria superiore, Vol. I, II, III, Lecture Notes. Istituto Matematico, Università di Roma.
—— 1967. Introduction to Galois geometries. Memorie Lincei (8) 7: 133-236.
—— 1973. Proprietà elementari relative ai segmenti ed alle coniche sopra un campo qualisiasi ed una congettura di Seppo Ilkka per il caso dei campi di Galois. Ann. Mat. Pura e Appl. ser. IV, XCVI: 289-337.
Tallini, G. (to appear) Graphic characterization of algebraic varieties in a Galois space. In: Atti Convegno Teorie Combinatorie, Acc. Lincei, Rome 1973.
Tallini Scafati, M. (to appear) The k-sets of type (m, n) in Galois spaces $S_{r,q}$ $(r \geq 2)$. In: Atti Convegno Teorie Combinatorie, Acc. Lincei, Rome 1973.
Thas, J.A. 1970. Connection between the n-dimensional affine space $A_{n,q}$ and the curve C, with equation $y = x^q$, of the affine plane A_{2,q^n}. Rend. Trieste 2: 146-151.
—— 1973. 4-gonal configurations. In: Finite Geometric Structures and Their Applications, C.I.M.E. II ciclo 1972, pp. 249-263. Cremonese, Rome.

—— 1973. On 4-gonal configurations. Geom. Dedicata 2: 317-326.

—— 1973. Construction of partial geometries. Simon Stevin 46: 95-98.

—— 1974. Construction of maximal arcs and partial geometries. Geom. Dedicata 3: 61-64.

—— (to appear) Some results concerning $\{(q + 1)(n - 1); n\}$-arcs and $\{(q + 1)(n - 1) + 1; n\}$-arcs in finite projective planes of order q.

—— (to appear) On 4-gonal configurations with parameters $r = q^2 + 1$ and $k = q + 1$.

CONFIGURATIONAL PROPOSITIONS IN PROJECTIVE SPACES

Judita Cofman

A configurational proposition about points and lines of a projective space Σ is any statement about points and lines in Σ of the following kind:

(C) For any choice of points $P_1, P_2, \ldots, P_n$ of Σ and any choice of lines $\ell_1, \ell_2, \ldots, \ell_m$ of Σ from certain inequalities $P_i \neq P_j$, $\ell_s \neq \ell_t$, certain incidences $P_i \; I \; \ell_s$, and negations of some incidences $P_i \not{I} \; \ell_s$, where $i, j \in \{1, 2, \ldots, n'\}$, $s, t \in \{1, 2, \ldots, m'\}$ with $n' \geq n$, $m' \geq m$, $i \neq j$, $s \neq t$, certain incidences of the same type follow.

The points $P_1, \ldots, P_n$ and the lines $\ell_1, \ldots, \ell_m$ are called the variables of (C); the points $P_{n+1}, \ldots, P_{n'}$ and the lines $\ell_{m+1}, \ldots, \ell_{m'}$ are the fixed elements of (C). If (C) has no fixed elements, we say that (C) is universally valid.

Well-known examples of configurational propositions are the proposition of Desargues and the proposition of Pappus. Their importance in the theory of projective spaces justifies the study of configurational propositions in general. It would be desirable to establish hierarchies between propositions of given classes of projective planes (see, for example, Amitsur [1]); however even special cases of this problem turn out to be extremely difficult. On the other hand there are other possibilities for the study of propositions by relating them to different properties of the corresponding spaces. For instance, R. Baer [3] has pointed out that (P, g)-transitivity of a projective plane π is equivalent to the property that π is (P, g)-desarguesian. This result has a large number of important consequences for projective planes (see, for instance, problems related to the classification of Lenz-Barlotti in [7], p. 126). Or questions of the following kind can be investigated: given a configurational proposition (C) in a restricted form in a projective space Σ, is it possible to remove some of the restrictions imposed on (C); in particular under which circumstances is (C) universally valid in Σ?

Here I would like to consider some recent results on configurational propositions. The following topics will be considered: (1) configurational propositions and correlations in projective spaces; (2) configurational pro-propositions in spreads; (3) configurational propositions in derivable planes.

I. CONFIGURATIONAL PROPOSITIONS AND CORRELATIONS

I.1. Projective planes

A correlation of a projective plane π is a 1-1 mapping of the point set $\mathcal{P}$ of π onto the line set $\mathcal{L}$ of π preserving the incidence relation in π. A point X (line x) is called an absolute point (line) of a correlation σ of π if and only if it is incident with its image X^σ (x^σ) under σ.

A correlation of order 2 is called a polarity. Although polarities of projective planes have often been investigated, not much is known about correlations of arbitrary power.

In this section we shall discuss results of Herzer [9] on correlations with a large number of absolute points

Correlations with many absolute points were studied by Baer [3] in 1942. Call a correlation δ of a projective plane π a (P, g)-correlation if and only if there is a non-incident point-line pair (P, g) in π such that $X^\delta = PX$ for any point X I g and $\ell^\delta = \ell \cap g$ for any line ℓ I P. It is easy to verify the following: (a) the points of g and the lines through P are the only absolute elements of δ; (b) for any point $Y \neq P$, $Y \not\mathrm{I}\ g$ and any line h I $PY \cap g$, $h \neq PY$, g there is at most one (P, g) correlation of π mapping Y onto h. A plane π is said to be (P, g)-homogeneous if and only if for a given non-incident point-line pair (P, g), any point $Y \neq P$, $Y \not\mathrm{I}\ g$ and any line ℓ I PY $\cap$ g, $\ell \neq$ PY, g there is a (P, g)-correlation of π mapping Y onto ℓ. Baer has proved that a plane π is pappian if and only if there are two distinct lines m, g in π such that π is (M, g)-homogeneous for any point M I m, M $\neq$ m $\cap$ g.

Consider now a correlation δ of a projective plane π with the property that the set of absolute points of δ contains the point set [g] of a line g of π as a proper subset. Then the following is true:

PROPOSITION 1.1. *If the set $\mathcal{A}$ of absolute points of a correlation δ of π contains the point set [g] of a line g of π as a proper subset, then there exists a unique line ℓ of π such that $\mathcal{A} = [g] \cup [\ell]$.*

PROOF

STEP 1. $A = g^{\delta^{-1}} \not\mathrm{I}\ g$; *all lines through* A *are absolute lines*. For, the image of any line $\ell \neq g$ through A is the point $A^{\delta} \cap (\ell \cap g)^{\delta}$. Since $(\ell \cap g)^{\delta} = \ell \cap g$, it follows that $\ell^{\delta} = \ell \cap g$. There are at least two distinct lines ℓ_1, ℓ_2 through A distinct from g. The points $\ell_1^{\delta} = \ell_1 \cap g$ and $\ell_2^{\delta} = \ell_2 \cap g$ are distinct. Hence A $\not\mathrm{I}$ g. All lines through A are absolute.

STEP 2. $g^{\delta} = B \neq A$. Otherwise δ would be an (A, g)-correlation, contradicting the existence of a point in $\mathcal{A}$- [g]. Clearly, B $\not\mathrm{I}$ g.

STEP 3. *No point of AB distinct from* P = AB $\cap$ g *is* an absolute point of δ. For, suppose Y is a point of AB such that $Y \neq P$. The images $A^\delta = g$, $P^\delta = AB$, and Y^δ are concurrent. Hence $Y^\delta \not I Y$.

STEP 4. For any point $X \in \mathcal{A} - [g]$, X^δ I A. The points A, X, $AX \cap g$ are three distinct points, hence their images $A^\delta = g$, X^δ I X, and $(AX \cap g)^\delta$ I $AX \cap g$ are three concurrent lines. Thus $X^\delta = AX$.

STEP 5. $\mathcal{A} - [g] = [\ell]$ for a line ℓ through P, different from g. Since for any point $X \in \mathcal{A} - [g]$ the line X^δ is incident with A it follows that $A^{\delta^{-1}}$ I X for every $X \in \mathcal{A} - [g]$. Take any point Z of $\ell = A^{\delta^{-1}}$, $Z \neq \ell \cap g$. It is easy to see that $Z^\delta = AZ$. Thus Z is an absolute point. Since every point of ℓ is absolute and since there are no absolute points on AB except P, it follows that g, ℓ, and AB are concurrent.

DEFINITION. A correlation δ is said to be a (A_i, g_i)-correlation (i = 1, 2) if and only if the set $\mathcal{A}$ of absolute points of δ consists of the points of g_i and $A_i = g_i^{\delta^{-1}}$ for i = 1, 2.

Our aim is to investigate projective planes admitting (A_i, g_i)-correlations. We shall consider the following questions: Which projective planes can admit (A_i, g_i)-correlations? How is it possible to characterize pappian projective planes in terms of (A_i, g_i)-correlations?

The following definitions are needed:

DEFINITION. An (A_i, g_i)-configuration consists of two distinct lines g_1, g_2 intersecting in a point P and of two distinct points A_1, A_2 incident with a line h through P where $A_1 \neq P \neq A_2$.

DEFINITION. For a given (A_i, g_i)-configuration consider any line $\ell \not{I}\, P$ intersecting g_i in R_i, $i = 1, 2$, and any point $S\ I\ \ell$, $S \neq \ell \cap h$, $R_1 \neq S \neq R_2$. If for all possible choices of ℓ and S the points $SA_1 \cap g_1$, $SA_2 \cap g_2$, and $R_1A_2 \cap R_2A_1$ are collinear we say that π is (A_i, g_i)-pappian.

THEOREM 1.1. <u>A projective plane</u> π <u>admits a</u> (A_i, g_i)-<u>correlation if and only if it is</u> (A_i, g_i)-<u>pappian</u>.

PROOF

(a) Assume that π is (A_i, g_i)-pappian. Denote the set of points in π which are not on h by $\bar{\mathscr{P}}$ and the set of lines in π not through P by $\bar{\mathscr{L}}$ Consider the following mapping $\bar{\sigma}$ of $\bar{\mathscr{P}}$ into $\bar{\mathscr{L}}$: for any $Y \in \bar{\mathscr{P}}$ let $Y^{\bar{\sigma}}$ be the line joining $YA_1 \cap g_1$ to $YA_2 \cap g_2$. It is easy to verify that $\bar{\sigma}$ is a bijection of $\bar{\mathscr{P}}$ onto $\bar{\mathscr{L}}$. Moreover, three collinear points of $\bar{\mathscr{P}}$ are mapped onto concurrent lines of $\bar{\mathscr{L}}$. Namely, if three points A, B, C are on a line $\ell \not{I}\, P$ then each of the lines $A^{\bar{\sigma}}$, $B^{\bar{\sigma}}$, $C^{\bar{\sigma}}$ is incident with the point $A_1R_2 \cap A_2R_1$ (where $R_i = \ell \cap g_i$, $i = 1, 2$) since π is (A_i, g_i)-pappian. If A, B, C are on a line t I P, consider the intersection

Q of $A^{\bar{\sigma}}$ and $B^{\bar{\sigma}}$. The point Q must belong to h; otherwise the lines through Q would represent the images of the points of $\bar{\mathscr{P}}$ on a line not through P under $\bar{\sigma}$, which is impossible. But if Q I h, this implies that $Q = A^{\bar{\sigma}} \cap B^{\bar{\sigma}}$ and similarly $Q = A^{\bar{\sigma}} \cap C^{\bar{\sigma}}$, since $Q = A^{\bar{\sigma}} \cap h$.

Thus $\bar{\sigma}$ is an isomorphism of the affine plane $\mathscr{A}$ obtained from π by deleting h together with its points onto the dual of the structure obtained from π by deleting P together with the lines through P. Obviously, $\bar{\sigma}$ can be uniquely extended to an isomorphism σ of the corresponding projective planes. Thus σ is a correlation of π. If $T \in g_i$ for i = 1 or 2, then $T^{\sigma} = (TA_1 \cap g_1)(TA_2 \cap g_2) = TA_{i+1}$ (i + 1 taken mod 2). Thus T I T^{σ}, in other words σ is a (A_i, g_i)-correlation.

(b) Let δ be a (A_i, g_i)-correlation of π. From the investigations in the proof of Proposition 1.1 it follows that $A_i \not{I}\, g_i$ and that all lines through A_i are absolute for i = 1, 2. Denote by h the line A_1A_2, by $\bar{\mathscr{P}}$ the point set of π outside h, and by $\bar{\mathscr{L}}$ the line set of π not through P. For any point $X \in \bar{\mathscr{P}}$ its image X^{δ} is the line through $XA_i \cap g_i$, i = 1, 2. The image ℓ^{δ} of any line $\ell \in \bar{\mathscr{L}}$ is the point $R_1A_2 \cap R_2A_1$ where $R_i = \ell \cap g_i$. The correlation δ is uniquely determined on the elements of $\bar{\mathscr{P}} \cup \bar{\mathscr{L}}$. As shown in (a) this implies that δ is uniquely determined on π. The points of g_i are mapped onto the lines through A_{i+1}. Thus $g_i^{\delta} = A_{i+1}$ for i = 1, 2, (i +1 taken mod 2). Hence

$(g_1 \cap g_1)^{\delta} = A_1A_2$ and $(A_1A_2)^{\delta} = g_1 \cap g_2$. In other words, δ interchanges h and P. The points A_1, A_2, P are collinear; they form together with g_1 and g_2 a (A_i, g_i)-configuration. Let ℓ be any line of π not through P and let S be any point of ℓ distinct from $\ell \cap h$ and from $R_i = \ell \cap g_i$, i = 1, 2. Since δ maps R_1, S, R_2 onto concurrent lines, it follows that π is (A_i, g_i)-pappian.

This proves Theorem 1.1.

REMARK 1.1. A (A_i, g_i)-correlation cannot be a polarity, since $A_i^{\delta^2} = A_{i+1}$ (i = 1, 2, i + 1 taken mod 2).

REMARK 1.2. A (A_i, g_i)-correlation is uniquely determined by A_i, g_i, i = 1, 2. Thus for a given (A_i, g_i)-configuration in π the plane π can admit at most one (A_i, g_i)-correlation.

DEFINITION. A projective plane π is said to be (A_1, g_1, g_2)-pappian if it is (A_i, g_i)-pappian for fixed elements A_1, g_1, g_2 and any point A_2 on $h = A_1P$ distinct from A_1 and P.

We are now able to prove the following characterization of pappian planes:

THEOREM 1.2. <u>A projective plane is pappian if and only if it is</u> (A_1, g_1, g_2)-<u>pappian</u>.

PROOF. We have only to show that (A_1, g_1, g_2)-pappian

implies that π is pappian. This will be done in several steps.

STEP 1. π is (A_1, g_2)-transitive. For, take two distinct points R, S on h, different from P and A_1. Denote by δ_R and δ_S the (A_1, R, g_1, g_2)- and (A_1, S, g_1, g_2)-correlations respectively. Obviously, $\delta_R\delta_S^{-1}$ is a collineation of π mapping R onto S. Moreover $\delta_R\delta_S^{-1}$ fixes every line through A_1 and every point on g_2. Thus $\delta_R\delta_S^{-1}$ is a (A_1, g_2)-homology mapping R onto S. By fixing R and varying S over the set of all points on h distinct from A_1 and P we obtain that π is (A_1, g_2)-transitive.

STEP 2. π is (A_1, h)-transitive. Take a point R on h distinct from P and A_1 and a line s I P, $s \neq g_1, g_2, h$. Denote by S the image of s under the inverse δ_R^{-1} of the (A_1, R, g_1, g_2)-correlation δ_R. It is easy to verify that $\delta_R^{-1}\ \delta_S\ \delta_R$ is the (A_1, R, g_1, s)-correlation of π. By fixing s and varying R over the points of h distinct from A_1 and P, we get (A_1, X, g_1, s)-correlations for all points X I h, $X \neq A_1$, P. Thus, in view of Step 1, the plane π is (A_1, s)-transitive. The (A_1, s)-transitivity and the (A_1, g_1)-transitivity together imply that π is (A_1, h)-transitive.

STEP 3. π is a translation plane with respect to h. Namely, the collineation δ^2 interchanges A_1 and A_2. This implies that π is also (A_2, h)-transitive. (A_1, h)-transitivity and (A_2, h)-transitivity imply that π is a translation plane with respect to h.

STEP 4. π is the dual of a translation plane with respect to P. This follows from the fact that δ interchanges h and P.

STEP 5. In the ternary ring $\mathcal{R}$ coordinatizing π with respect to the quadrangle $O = (o, o)$, $E = (1, 1)$, $V = (\infty)$, $U = (o)$ by the method of Hall [8], with $V = P$, $VO = g_1$, $VE = g_2$, $U = A_1$, and $A_2 = (a)$ for any element $a \in \mathcal{R} - \{o\}$ the following relation is satisfied:

(*) $(a - ax)z = az - (az)x$ for all $x, z \in \mathcal{R}$.

For, consider the incident point-line pair

$Q = (x, y)$, $g = [az, o]$ with $y = (az)x$.

Then $Q^{\delta} = ([(x, y), (o)] \cap OV)\ ([(x, y), (a)] \cap EV)$

$= [a - ax, y]$

and $g^{\delta} = [([az, o] \cap OV), (a)] \cap [[az, o] \cap EV, (o)]$

$= (z, az)$

(using the rules given in the proof of Theorem 1.2 (b)).

Since $Q^{\delta}\ I\ g^{\delta}$, it follows that

$az = (a - ax)z + y = (a - ax)z + (az)x$,

or

(*) $(a - ax)z = az - (az)x$ for all $x, z \in \mathcal{R}$.

STEP 6. $\mathcal{R}$ is a commutative field. π is a translation plane with respect to h and the dual of a translation plane with respect to V ; thus $\mathcal{R}$ is first of all a distributive quasifield. Moreover, the plane π is (A_1, g_2)-transitive and since π is a translation plane with respect to PA_1, it follows that π is also (A_1, g_1)-transitive. This implies that multiplication is associative in $\mathcal{R}$ In view of Step 5, relation (*) is valid for any $a \in \mathcal{R} - \{o\}$. Take a = 1; this implies that xz = zx for all x, z in $\mathcal{R}$. The ternary ring $\mathcal{R}$ is commutative.

In other words, we have proved that π is pappian.

REMARK 1.3. As we have shown in Step 1, if π is (A_1, g_1, g_2)-pappian then π is (A_1, g_2)-desarguesian. Using the same arguments as in Step 1, one can prove that if π is pappian, then π is (A, g)-desarguesian for all non-incident point-line pairs in π. This implies that π is (X, g)-desarguesian for all point-line pairs in π, i.e. π is desarguesian. Thus we have a new proof of the well-known statement: a pappian plane is desarguesian.

Our next aim is to give a sufficient condition for the existence of a (A_i, g_i)-correlation in a projective plane:

THEOREM 1.3. <u>Let</u> $\mathcal{R}$ <u>be the ternary ring</u>[1] <u>of a projective plane</u> π <u>with respect to the points of reference</u>: O, E, U, V. <u>If</u> π <u>is</u> (V, UV)-<u>transitive and if the relation</u>

(*) $(a - ax)z = az - (az)x$

<u>is satisfied for all</u> $x, z \in \mathcal{R}$, <u>then</u> π <u>admits a</u> (A_i, g_i)-<u>correlation with</u> $A_1 = (o)$, $A_2 = (a)$, $g_1 = OV$, $g_2 = EV$ <u>for</u>

1. Throughout the whole section the coordinatization method of Hall [8] is used.

a fixed element $a \in \mathcal{R} - \{o\}$.

PROOF. Denote by $\bar{\mathcal{P}}$ the set of the points (x, y) in π and by $\bar{\mathcal{L}}$ the set of all lines $[m, k]$ in π. Define the mapping $\bar{\delta}$ on the set $\bar{\mathcal{P}} \cup \bar{\mathcal{L}}$ as follows:

$$(x, y)^{\bar{\delta}} = [a - ax, y]$$

and $[az, k]^{\bar{\delta}} = (z, az + k)$.

It remains as an exercise to verify that $\bar{\delta}$ is an isomorphism of the affine plane $\mathcal{A}$ obtained from π by deleting UV with its points onto the affine plane $\mathcal{B}$ which is the dual of the structure obtained from π by deleting P together with all lines through P. Hence $\bar{\delta}$ can be extended uniquely to a correlation δ of π. It is easy to check that the points of VO and VE are absolute points of δ; the points A_1 and (a) play the role of A_1 and A_2. This proves Theorem 1.3.

COROLLARY. There exist non-desarguesian projective planes admitting (A_i, g_i)-correlations.

PROOF. There are known examples of distributive quasi-fields with commutative multiplication which are not fields (see, for example, Dembowski [7], 5.3). All of these satisfy relation (*) with $a = 1$ and their corresponding projective planes are (V, UV)-transitive. Thus any projective plane coordinatized by a distributive quasi-field with commutative multiplication admits a (A_i, g_i)-correlation.

REMARK 1.4. It would be interesting to know which classes of the Lenz classification for projective planes can admit (A_i, g_i)-correlations. (For the definition of the Lenz classification see Dembowski [7].) I shall mention without proof the following result of Herzer ([9], Satz 8):

THEOREM 1.4. <u>If a projective plane</u> π <u>is</u> (X, y)-<u>transitive for at least one incidence point-line pair</u> (X, y) <u>and if</u> π <u>admits an</u> (A_i, g_i)-<u>correlation</u>, <u>then</u> π <u>must belong to one of the following classes</u>: <u>Lenz class II</u>, <u>Lenz class III</u>, <u>or Lenz class V</u>.

REMARK 1.5. It is not known whether there exist planes of Lenz class III admitting (A_i, g_i)-correlations. Planes of Lenz class II admitting (A_i, g_i)-correlations have been constructed by Herzer ([9], p. 251). Distributive quasi-fields with commutative multiplication provide examples for planes of class V with (A_i, g_i)-correlations.

I.2. Desarguesian projective spaces

For an integer $n \geq 3$ and a not necessarily commutative field K denote the n-dimensional left vector space over by $V(n,K)$. Let the projective space isomorphic to the lattice $\mathscr{L}$ of the subspaces of $V(n,K)$ be denoted by Σ. A correlation of Σ is an anti-automorphism of $\mathscr{L}$. A subspace Λ of Σ is called totally isotropic, isotropic, or non-isotropic with respect to a correlation of Σ if and only if $\Lambda \cap \Lambda^{\delta} = \Lambda$, or $\Lambda \cap \Lambda^{\delta} \neq \phi$, or $\Lambda \cap \Lambda^{\delta} = \phi$ respectively. Isotropic points are totally isotropic; isotropic points will be called absolute points of δ.

Herzer [10] has generalized the notion of (A_i, g_i)-correlations for desarguesian spaces of arbitrary finite dimension $n \geq 3$:

DEFINITION. A correlation δ of Σ is called a $\mathcal{G}$-correlation if and only if there exists a set $\mathcal{G}$ of $n - 1$ non-isotropic lines $g_1, \ldots, g_{n-1}$ of δ in Σ with the following properties: (1) all points of g_i, $i = 1, \ldots, n - 1$, are absolute points of δ; (2) $\bigcap_{i=1}^{n-1} g_i = P$; (3) the projective space $\langle g_i;\ i = 1, \ldots, n - 1\rangle$ spanned by $g_1, \ldots, g_{n-1}$ is the is the whole space Σ.

We are able to prove the following theorem of Herzer:

THEOREM 1.5. <u>Σ admits a $\mathcal{G}$-correlation if and only if Σ is pappian</u>.

PROOF

(a) Suppose that Σ admits a $\mathcal{G}$-correlation. Take any line $g_i \in \mathcal{G}$. Let X be an arbitrary point of g_i. Since X is an absolute point of δ, the hyperplane X^{δ} of Σ is incident with X. Furthermore $X^{\delta} \cap g_i = X$, since g_i is non-isotropic. (Otherwise g_i, as a line of X^{δ}, would intersect g_i^{δ} which is contained as a hyperplane in X^{δ}.) Thus the mapping $\mu: X \to X^{\delta} \cap g_i$ is the identity map of the point set $[g_i]$ of g_i onto itself. Thus μ is a projectivity, and hence δ is a projective correlation (see, for example, Lenz [13]). It is well known that if Σ admits a projective correlation then Σ is pappian.

(b) Let Σ be a pappian space. In that case $\mathcal{K}$ is commutative and we are able to construct a $\mathcal{G}$-correlation δ by using the following bilinear form $\mathbb{B}$ over a given basis $\vec{v}_1, \ldots, \vec{v}_n$ of V: $(\vec{v}_i, \vec{v}_j) = a_{ij}$ where $a_{ii} = o$ for $i = 1, \ldots, n$, $o \neq a_{in} = -a_{ni}$ for $i = 1, 2, \ldots, n - 1$, and let $((a_{ij})) \neq o$. It is easy to verify that the correlation δ induced by $\mathbb{B}$ is a $\mathcal{G}$-correlation with respect to $g_i \in \mathcal{G}$, $i = 1, \ldots, n - 1$, where g_i is the line of Σ obtained from the 2-dimensional subspace $\langle g_i, g_n \rangle$ of V.

REMARK 1.6. A $\mathcal{G}$-correlation can be a polarity. According to Herzer ([13], Satz 6), in a projective space Σ the set of all $\mathcal{G}$-correlations which are polarities is the set of all symplectic polarities of the space.

II. CONFIGURATIONAL PROPOSITIONS IN SPREADS

II.1

A set $\mathcal{S}$ of subspaces of the same dimension in a projective space Σ is called a spread of Σ if and only if every point of Σ is contained in exactly one element of $\mathcal{S}$. The spread $\mathcal{S}$ is called a spread over $\mathcal{K}$ if $\mathcal{K}$ is the not necessarily commutative field coordinatizing Σ. Throughout the whole of section II we shall consider only spreads of Σ with the following property ($\mathbb{P}$): any two distinct elements of $\mathcal{S}$ span Σ. From any spread $\mathcal{S}$ with property ($\mathbb{P}$) a translation plane π can be obtained in the following way. Embed Σ as a hyperplane in a projective space Σ'. The points of Σ' not contained in Σ and the subspaces of Σ' intersecting Σ in the elements of $\mathcal{S}$ are the points and lines, respectively, of an affine plane π, where incidence is the incidence of Σ' restricted to the elements of π. Every elation of Σ' with axis Σ induces a translation of π. It follows easily that π is a translation plane. Our aim is to investigate the following problem: how does the structure of a spread $\mathcal{S}$ influence the structure of the corresponding translation plane π?

DEFINITION. Any line ℓ meeting three distinct elements $\mathcal{A}, \mathcal{B}, \mathcal{C}$ of a spread $\mathcal{S}$ is called a $(\mathcal{A}, \mathcal{B}, \mathcal{C})$-transversal.

DEFINITION. A spread $\mathcal{S}$ is called regular if and only if for any four distinct elements $\mathcal{A}, \mathcal{B}, \mathcal{C}, \mathcal{D}$ of $\mathcal{S}$ the element $\mathcal{D}$ is met either by all $(\mathcal{A}, \mathcal{B}, \mathcal{C})$-transversals or by none of them. The elements of a regular spread intersected by all $(\mathcal{A}, \mathcal{B}, \mathcal{C})$-transversals form the regulus $\mathcal{R} = \mathcal{R}(\mathcal{A}, \mathcal{B} \cdot \mathcal{C})$.

DEFINITION. A spread is called Moufang, desarguesian,

or pappian if and only if the corresponding translation plane is Moufang, desarguesian, or pappian respectively.

Bruck and Bose [5] have investigated spreads $\bar{\mathcal{S}}$ in finite-dimensional projective spaces Σ over fields with at least three elements. About these spreads $\bar{\mathcal{S}}$ they have obtained the following results:

(1) Any pappian spread $\bar{\mathcal{S}}$ is regular.

(2) Any regular spread $\bar{\mathcal{S}}$ is a Moufang spread.

Herzer [11] has observed that the same results are valid for spreads in projective spaces Σ of arbitrary dimensions over fields with at least three elements. Furthermore, Herzer has obtained characterizations of pappian, desarguesian, and Moufang spreads in terms of certain configurational propositions. In II.2, we shall investigate the configurational propositions given by Herzer and, in II.3, we shall outline the proofs of his main results.

II.2. The propositions (D) and (P)

Let $\mathcal{S}$ be a regular spread of a projective space Σ. For any element $\mathcal{A} \in \mathcal{S}$, call any line of Σ intersecting $\mathcal{A}$ in exactly one point a transversal of $\mathcal{A}$.

In the set $\bar{T}_{\mathcal{A}}$ of all transversals of $\mathcal{A}$ we define a binary operation "$\underset{\mathcal{A}}{||}$" called $\mathcal{A}$-parallelism as follows: two arbitrary lines a, b of $\bar{T}_{\mathcal{A}}$ are called $\mathcal{A}$-parallel ($a \underset{\mathcal{A}}{||} b$) if and only if a and b are transversals of a common regulus of $\mathcal{S}$. Obviously, $\mathcal{A}$-parallelism is an equivalence relation in $\bar{T}_{\mathcal{A}}$. Moreover, for any $a \in \bar{T}_{\mathcal{A}}$ and any point $X \in \mathcal{A}$ there is exactly one line through X which is $\mathcal{A}$-parallel to a.

We are now able to define two configurational propositions (D) and (P) for regular spreads.

DEFINITION. Let $\mathcal{A}, \mathcal{B}, \mathcal{C}$ be three distinct elements of a regular spread $\mathcal{S}$ and let A_i, B_i, C_i, i = 1, 2, be points of the projective space Σ such that $A_i \in \mathcal{A}$, $B_i \in \mathcal{B}$, and $C_i \in \mathcal{C}$. We say that A_i, B_i, C_i satisfy the configurational proposition (D) if and only if $A_1B_1 \parallel_{\mathcal{A}} A_2B_2$ and $A_1C_1 \parallel_{\mathcal{A}} A_2C_2$ imply $B_1C_1 \parallel_{\mathcal{A}} B_2C_2$.

DEFINITION. The configurational proposition (D) is universally valid in $\mathcal{S}$ if and only if (D) is true for any three distinct elements $\mathcal{A}, \mathcal{B}, \mathcal{C}$ of $\mathcal{S}$ and for any possible choice of $A_i \in \mathcal{A}$, $B_i \in \mathcal{B}$, and $C_i \in \mathcal{C}$.

DEFINITION. Let $\mathcal{A}, \mathcal{B}$ be two distinct elements of a regular spread $\mathcal{S}$ of Σ and let A_1, A_2, A_3 be points of $\mathcal{A}$ and B_1, B_2, B_3 be points of $\mathcal{B}$. We say that A_1, A_2, A_3, B_1, B_2, B_3 satisfy the configurational proposition (P) if and only if from $A_1B_2 \parallel_{\mathcal{A}} B_1A_2$ and $A_2B_3 \parallel_{\mathcal{A}} A_3B_2$ it follows that

$A_3B_1 \parallel_{\mathcal{A}} A_1B_3$.

DEFINITION. In a regular spread Proposition (P) is <u>universally valid</u> if and only if it is valid for any two distinct elements $\mathcal{A}$, $\mathcal{B}$ of $\mathcal{S}$ and all possible choices of $A_i \in \mathcal{A}$, $B_i \in \mathcal{B}$, $i = 1, 2, 3$.

The universal validity of (D) or (P) in $\mathcal{S}$ is connected with the existence of collineation groups induced on the elements of $\mathcal{S}$:

LEMMA 2.1. <u>Let $\mathcal{S}$ be a regular spread of Σ. Proposition (D) is universally valid in $\mathcal{S}$ if and only if Σ admits a collineation group Δ which fixes $\mathcal{S}$ elementwise and is transitive on the points of any element of $\mathcal{S}$</u>.

PROOF. Assume that (D) is universally valid in $\mathcal{S}$. Take an arbitrary but fixed element $\mathcal{A} \in \mathcal{S}$. Let O and P be two points of $\mathcal{S}$. On the set $\bar{\mathcal{P}}$ of points of Σ not contained in $\mathcal{A}$ define the map $\bar{\delta}_{OP}$: $X \to \ell_P \cap \mathcal{X}$ where ℓ_P is the $\mathcal{A}$-parallel to OX through the point P and $\mathcal{X}$ the element of $\mathcal{S}$ through X. It is easy to verify that $\bar{\delta}_{OP}$ is a 1-1 map of $\bar{\mathcal{P}}$ onto itself. Our aim is to show that $\bar{\delta}_{OP}$ can be uniquely extended to a collineation δ_{OP} of Σ. For this we observe the following: (1) Points of $\bar{\mathcal{P}}$ on a transversal of any $\mathcal{B} \in \mathcal{S}$ are mapped onto points of a transversal in view of (D). (2) Points of $\bar{\mathcal{P}}$ on a line $h \in \mathcal{B}$ for any $\mathcal{B} \in \mathcal{S}$ are mapped onto collinear points. This follows from the fact that h can be embedded into a plane π not contained in any element of $\mathcal{S}$; in view of (1) $\pi^{\bar{\delta}_{OP}}$ is a plane. (3) Coplanar points of Σ can be embedded into three-dimensional spaces not contained in any element of $\mathcal{S}$; hence coplanar points of Σ are mapped onto coplanar points. Thus $\bar{\delta}_{OP}$ can be extended to a map δ_{OP} of the point set $\mathcal{P}$ of Σ into itself in such a way that δ_{OP} is a collineation of Σ. Clearly, δ_{OP} fixes every element of $\mathcal{S}$. Denote by Δ the set of all collineations δ_{OX} for a fixed point $O \in \mathcal{A}$ and all points $X \in \mathcal{A}$. The set Δ is a group acting as a transitive permutation group on the point set of every $\mathcal{B} \in \mathcal{S}$.

Conversely, if Δ is a collineation group of Σ with the required properties, then for any $\mathcal{B}$-transversal g of $\mathcal{S}$ and any $\delta \in \Delta$, $g^{\delta} \underset{\mathcal{B}}{||} g$. The group Δ is transitive on the points of $\mathcal{B}$. Hence (D) is universally valid in $\mathcal{S}$.

LEMMA 2.2. Let $\mathcal{A}$, $\mathcal{B}$ be two distinct elements of the regular spread $\mathcal{S}$ and let g be a line intersecting both $\mathcal{A}$ and $\mathcal{B}$. Define the map δ_g on the point set of $\mathcal{A}$ and $\mathcal{B}$ as follows: $X^{\delta_g} = \ell_{X,g} \cap \mathcal{B}$ for any $X \in \mathcal{A}$ and $X^{\delta_g} = \ell_{X,g} \cap \mathcal{A}$ for any $X \in \mathcal{B}$. Here $\ell_{X,g}$ denotes the $\mathcal{A}$-parallel to g through X. Proposition (P) is universally valid in $\mathcal{S}$ if and only if for any two distinct elements $\mathcal{A}$, $\mathcal{B}$ of $\mathcal{S}$ and any three lines a, b, c intersecting $\mathcal{A}$ and $\mathcal{B}$ the following relation is true:

(*) $\quad (\delta_a\delta_b\delta_c)^2 = 1.$

PROOF. Suppose that (P) holds. If any two of the lines

a, b, c coincide, then (*) is satisfied, since $\delta_a^2 = \delta_b^2 = \delta_c^2 = 1$. Otherwise take any point $X \in \mathcal{A}$ and consider the configuration $A_1 = X$, $A_2 = X^{\delta_a\delta_b\delta_c\delta_a}$, $A_3 = X^{\delta_a\delta_b}$, $B_1 = X^{\delta_a\delta_b\delta_c}$, $B_2 = X^{\delta_a}$, and $B_3 = X^{\delta_a\delta_b\delta_c\delta_a\delta_b}$. Because of (P) it follows that $X = A_1 = B_3{}^{\delta_c} = X^{(\delta_a\delta_b\delta_c)^2}$. Thus $(\delta_a\delta_b\delta_c)^2 = 1$.

Conversely, by assuming that (*) holds for the above configuration A_1, A_2, A_3, B_1, B_2, B_3, we deduce that $A_1B_3 || A_3B_1$, i.e. (P) is satisfied.

LEMMA 2.3. <u>Let $\mathcal{S}$ be a regular spread in which (P) is universally valid. Let $\mathcal{A}$, $\mathcal{B}$ be two distinct elements of $\mathcal{S}$, g a fixed line meeting $\mathcal{A}$ and $\mathcal{B}$ and h any line of Σ meeting $\mathcal{A}$ and $\mathcal{B}$. Then the set $\Delta_g = \{\delta_g \cdot \delta_h \mid$ for all h meeting $\mathcal{A}$ and $\mathcal{B}\}$ induces on the projective space $\mathcal{A}$ an abelian collineation group transitive on the points of $\mathcal{A}$.</u>

PROOF. Let h and h' be two arbitrary lines intersecting $\mathcal{A}$ and $\mathcal{B}$. According to Lemma 2.2, $\delta_h\delta_g\delta_{h'} = (\delta_h\delta_g\delta_{h'})^{-1} = (\delta_{h'})^{-1}\delta_g^{-1}\delta_h^{-1} = \delta_{h'}\delta_g\delta_h$. Hence $(\delta_g\delta_h)(\delta_g\delta_{h'}) = \delta_g(\delta_h\delta_g\delta_{h'}) = \delta_g(\delta_{h'}\delta_g\delta_h) = (\delta_g\delta_{h'})(\delta_g\delta_h)$. Therefore Δ_g induces a commutative permutation group on the point set of $\mathcal{A}$. Let X and Y be two arbitrary points of $\mathcal{A}$. The points X^{δ_g} and Y determine a unique line h such that $Y = X^{\delta_g\delta_h}$; hence Δ_g is transitive on the points of $\mathcal{A}$. From the definition of δ_g and δ_h it follows that Δ_g induces a collineation group on $\mathcal{A}$, which proves the lemma.

II.3. Main results

We are now able to prove the following theorems:

THEOREM 2.1. *A regular spread $\mathcal{S}$ is desarguesian if and only if (D) is universally valid in $\mathcal{S}$.*

THEOREM 2.2. *A spread $\mathcal{S}$ is pappian if and only if $\mathcal{S}$ is regular and (P) is universally valid in $\mathcal{S}$.*

THEOREM 2.3. *Let $\mathcal{S}$ be a spread over a field K with $|K| > 2$. Then $\mathcal{S}$ is a Moufang spread which is not desarguesian if and only if $\mathcal{S}$ is regular and (D) is not universal in $\mathcal{S}$.*

For the proofs of these results some well-known properties of spreads are needed (see André [2] and Bruck and Bose [4], [5]):

PROPERTY 1 (Bruck-Bose [5], p. 155). Let $\mathcal{S}$ be a spread of Σ. Denote by V the vector space over K from which Σ is obtained. For any element $\mathcal{X} \in \mathcal{S}$ denote by $\bar{\mathcal{X}}$ the corresponding subspace of V. Take an ordered triple $\mathcal{A}, \mathcal{B}, \mathcal{C}$ of distinct elements of $\mathcal{S}$; there is a unique linear transformation $\vec{a} \to \vec{a}'$ of $\bar{\mathcal{A}}$ into $\bar{\mathcal{B}}$ such that the linear transformation $\vec{a} \to \vec{a} + \vec{a}'$ maps $\bar{\mathcal{A}}$ into $\bar{\mathcal{C}}$. The vectors of any element $\bar{\mathcal{X}}$ of the space corresponding to $\mathcal{X} \in \mathcal{S} - \mathcal{A}$ can be expressed in the form $\vec{a}\varphi_{\bar{\mathcal{X}}} + \vec{a}'$ where $\vec{a} \in \mathcal{A}$, $\vec{a}'$ is the corresponding vector of $\mathcal{B}$ defined above, and $\varphi_{\bar{\mathcal{X}}}$ is a unique linear transformation of $\bar{\mathcal{A}}$ into itself determined by $\bar{\mathcal{X}}$. In particular $\varphi_{\bar{\mathcal{B}}} = 0$ and $\varphi_{\bar{\mathcal{C}}} = I$. The collection $\mathbb{C}$ of the linear transformations $\varphi_{\bar{\mathcal{X}}}$ for all $\mathcal{X} \in \mathcal{S} - \mathcal{A}$ has the properties: $\varphi_{\bar{\mathcal{X}}} - \varphi_{\bar{\mathcal{Y}}}$ is non-singular if $\bar{\mathcal{X}} \neq \bar{\mathcal{Y}}$; if $\vec{a}$, $\vec{b} \in \mathcal{A}$ and $\vec{a} \neq \vec{o}$, then there exists exactly one element $\varphi_{\bar{\mathcal{X}}} \in \mathbb{C}$ such that $\vec{a}\varphi_{\bar{\mathcal{X}}} = \vec{b}$.

PROPERTY 2 (Bruck-Bose [5], p. 158). Let Σ' be a projective space containing Σ as a hyperplane and let π be the translation plane obtained from Σ', Σ, and $\mathcal{S}$ as described in II.1. Then there is a ternary ring of π isomorphic to the algebraic structure $\bar{\mathcal{A}}(+, \cdot)$ defined as follows: (1) the elements of $\bar{\mathcal{A}}(+, \cdot)$ are the vectors of $\bar{\mathcal{A}}$; (2) "+" is the vector addition; (3) let $\vec{e}$ be a fixed non-zero vector of $\bar{\mathcal{A}}$; then for any two vectors $\vec{a}$, $\vec{b} \in \bar{\mathcal{A}}$ the product $\vec{a} \cdot \vec{b}$ is the vector $\vec{a}\,\varphi_{\vec{b}}$ where $\varphi_{\vec{b}}$ is the transformation of $\mathbb{C}$ mapping $\vec{e}$ onto $\vec{b}$. Clearly $\vec{e}$ plays the role of the identity. $\bar{\mathcal{A}}(+, \cdot)$ is a quasifield.

PROPERTY 3 (Bruck-Bose [5], p. 159). If $\mathbb{C}$ is closed under multiplication, then $(\mathbb{C}, \cdot)$ is a group with zero isomorphic to $\bar{\mathcal{A}}(\cdot)$.

PROPERTY 4 (Herzer [11]). $\mathcal{K}$ is contained in the kernel of $\bar{\mathcal{A}}(+, \cdot)$. Moreover, $\mathcal{K}$ is contained in the centre of $\bar{\mathcal{A}}(+, \cdot)$ if and only if $\mathcal{S}$ is a regular spread.

PROPERTY 5 (André [2]). Let Q be a quasifield coordinatizing a translation plane π ; then the multiplicative group of the kernel of Q is isomorphic to the group of all homologies of π with a given affine centre and the improper line as axis.

We shall now indicate the proofs of Theorem 2.1, 2.2, and 2.3.

PROOF OF THEOREM 2.1. Suppose that (D) is universally valid in a regular spread $\mathcal{S}$. Then, according to Lemma 2.1 the space Σ admits the collineation group Δ. Denote by $\hat{\mathcal{K}}^*$ the multiplicative group of the kernel of $\bar{\mathcal{A}}(+, \cdot)$ and by $\mathcal{K}^*$ the multiplicative group of $\mathcal{K}$. It is not difficult to see that $\Delta \simeq \hat{\mathcal{K}}^*/\mathcal{K}^*$. On the other hand, if Δ^* denotes the group of homologies of π with improper axis and a common affine centre, then according to Property 5, $\Delta^* \simeq \hat{\mathcal{K}}^*$. Let $\langle\vec{r}\rangle$ be the 1-dimensional subspace of V' corresponding to an affine point R of π. Then the vectors $\langle\vec{r} + (\vec{x}, \vec{y})\rangle$ for $\vec{x}, \vec{y} \in \bar{\mathcal{A}}$ represent the affine points of π (V is represented in the form $\bar{\mathcal{A}} \oplus \bar{\mathcal{A}}$). Let $\langle\vec{r} + (\vec{x}, \vec{y})\rangle$ and $\langle\vec{r} + (\vec{x}', \vec{y}')\rangle$ be any two points of π collinear with $\langle\vec{r}\rangle$. Then $\langle(\vec{x}, \vec{y})\rangle$ and $\langle(\vec{x}', \vec{y}')\rangle$ are points of a common element $\mathcal{X} \in \mathcal{S}$. Since Δ is transitive on the points of $\mathcal{X}$, there is an element $\delta \in \Delta$ such that $\langle(\vec{x}', \vec{y}')\rangle = \langle(\vec{x}, \vec{y})\rangle\delta$. The isomorphisms $\Delta \simeq \hat{\mathcal{K}}^*/\mathcal{K}^*$ and $\hat{\mathcal{K}}^* \simeq \Delta^*$ imply the existence of a homology mapping $\langle\vec{r} + (\vec{x}, \vec{y})\rangle$ onto $\langle\vec{r} + (\vec{x}', \vec{y}')\rangle$. Hence π is (R, ℓ_∞)-transitive, i.e. π is desarguesian (see Pickert [14]).

Conversely, if π is desarguesian, then π is (R, ℓ_∞)-transitive. Hence Σ admits a collineation group Δ fixing all elements of $\mathcal{S}$ and acting transitively on the points of each element in $\mathcal{S}$ (because of $\Delta \simeq \hat{\mathcal{K}}^*/\mathcal{K}^*$, $\hat{\mathcal{K}}^* \simeq \Delta^*$). In view of Lemma 2.1 Proposition (D) is universally valid in $\mathcal{S}$.

PROOF OF THEOREM 2.2. Suppose $\mathcal{S}$ is pappian. Then the multiplicative group of $\bar{\mathcal{A}}(+, \cdot)$ is commutative. In view of $\Delta \simeq \hat{\mathcal{K}}^*/\mathcal{K}^*$ this implies that Δ is abelian. Let $\mathcal{A}$, $\mathcal{B}$ be two arbitrary elements of $\mathcal{S}$ with $A_1, A_2, A_3 \in \mathcal{A}$ and B_1, B_2, $B_3 \in \mathcal{B}$ such that $A_1B_2 \underset{\mathcal{A}}{||} A_2B_1$ and $A_2B_3 \underset{\mathcal{A}}{||} B_2A_3$. Let δ_2 be the element of Δ mapping A_1 onto A_2 and δ_3 be the element of Δ mapping A_2 onto A_3. Then $(A_1B_2)^{\delta_2} = A_2B_1$ and $(A_2B_3)^{\delta_3} = A_3B_2$. Hence $B_2^{\delta_2} = B_1$ and $B_3^{\delta_3} = B_2$.

The lines A_3B_1 and A_1B_3 are $\mathcal{A}$-parallel if and only if $(A_1B_3)^{\delta_2\delta_3} = A_3B_1$, i.e. if and only if $B_3^{\delta_2\delta_3} = B_1$. However, Δ is abelian, hence $B_3^{\delta_2\delta_3} = B_3^{\delta_3\delta_2} = B_1$. Thus (P) is universally valid in $\mathcal{S}$. Moreover, since $\mathcal{S}$ is pappian $\mathcal{S}$ is regular.

Conversely, suppose that (P) is universally valid in $\mathcal{S}$ and that $\mathcal{S}$ is regular. Then, in view of Lemma 2.3 the group Δ_g induced on $\mathcal{A}$ is abelian. Choose $\mathcal{A}, \mathcal{B}$, $\mathcal{C}$ as in

Property 1. If g is a $(\mathcal{A}, \mathcal{B}, \mathcal{C})$-transversal and h a $(\mathcal{A}, \mathcal{B}, \mathcal{D})$-transversal for any element $\mathcal{D} \varepsilon \mathcal{S} - \{\mathcal{A}\}$ then the collineation $\delta_g \delta_h$ of $\mathcal{A}$ is induced by the linear map $\varphi_{\overline{\mathcal{D}}} \varepsilon \mathbb{C}$ (Property 1). Thus $\Delta_g \simeq \mathbb{C}^*/\mathcal{K}^*$ where $\mathbb{C}^* = \mathbb{C} - \{o\}$.

Since $\mathcal{S}$ is regular, $\mathcal{K}^* \subseteq \mathbb{C}^*$, therefore $\mathbb{C}$ is closed under multiplication. Hence, according to Property 3, $\mathbb{C}^*(\cdot) \simeq \overline{\mathcal{A}}^*(\cdot)$. In view of Property 4 the group $\mathcal{K}^*$ is contained in the centre of $\overline{\mathcal{A}}^*$. Since $\overline{\mathcal{A}}^*/\mathcal{K}^* \simeq \mathbb{C}^*/_{\mathcal{K}}{}^* \simeq \Delta_g$, where Δ_g is abelian, this implies that $\overline{\mathcal{A}}^*(\cdot)$ is abelian. Therefore $\overline{\mathcal{A}}(+, \cdot)$ is a commutative field. Thus $\mathcal{S}$ is pappian.

PROOF OF THEOREM 2.3. If $\mathcal{S}$ is a regular spread over a field $\mathcal{K}$ with $|\mathcal{K}| > 2$ then $\mathcal{S}$ is at least a Moufang spread according to Bruck and Bose [5]. Proposition (D) is not universally valid in $\mathcal{S}$; hence according to Theorem 2.1 the spread $\mathcal{S}$ is not desarguesian.

Conversely, let $\mathcal{S}$ be a Moufang spread, which is not desarguesian. Then the quasifield $\overline{\mathcal{A}}(+, \cdot)$ (see Property 2) is a proper alternative ring. Hence the centre of $\overline{\mathcal{A}}(+, \cdot)$ coincides with its kernel (see Pickert [14]). Since the kernel of $\overline{\mathcal{A}}(+, \cdot)$ contains the field $\mathcal{K}$, this implies that $\mathcal{K}$ is contained in the centre of $\overline{\mathcal{A}}(+, \cdot)$. By Property 4, the spread $\mathcal{S}$ is a regular spread. Clearly (D) is not universally valid in $\mathcal{S}$.

III. CONFIGURATIONAL PROPOSITIONS IN DERIVABLE PROJECTIVE PLANES

III.1 Introduction

A projective plane π is called derivable if and only if it contains a derivation set $\mathcal{D}$, that is a set $\mathcal{D}$ of points on a line ℓ of π with the following property: for any two distinct points A, B of π not on ℓ for which the line AB intersects ℓ in a point of $\mathcal{D}$ there is a Baer subplane π_0 of π through A and B meeting ℓ exactly in the points of $\mathcal{D}$. The subplane π_0 is uniquely determined by A, B, and $\mathcal{D}$; if a Baer subplane π_0 of π contains the derivation set $\mathcal{D}$, then we say that π_0 belongs to $\mathcal{D}$.

The importance of derivable planes is underlined by the fact that from these planes new projective planes can be constructed (see Ostrom [12]). Various classes of derivable planes have been studied in detail; on the other hand it seems desirable to investigate properties common to all derivable planes. One of the first attempts in this direction was made by Prohaska [15], who has proved that every Baer subplane belonging to a derivation set of a finite derivable plane is desarguesian.

In this section derivable planes of arbitrary (finite or infinite) order are considered. A generalization of Prohaska's result is obtained by showing that any Baer subplane π_0 which belongs to a derivation set $\mathcal{D}$ of a derivable projective plane π is an affine hyperplane of a three-dimensional affine space $\mathcal{T}$ contained in π. Hence π_0 is desarguesian. Moreover any Baer collineation of π fixing π_0 pointwise induces a collineation of $\mathcal{T}$. This gives the following information about the group $\Pi(\pi_0)$ of all Baer collineations of π fixing π_0 pointwise: $\Pi(\pi_0)$ is a subgroup of the group of all perspectivities of $\mathcal{T}$ with axis π_0 and centres on an improper line of $\mathcal{T}$ not contained in π_0.

Our investigations are carried out on the dual planes of derivable planes; the Baer collineations and Baer subplanes belonging to a "dual derivation set" in these planes have the above-mentioned properties of π_0 and $\Pi(\pi_0)$.

III.2. Main results

Throughout this section π denotes a derivable projective plane, $\mathcal{D}$ a derivation set in π, and π_0 a Baer subplane belonging to $\mathcal{D}$.

Let us denote by $\hat{\pi}$ the plane dual to π. The dual of $\mathcal{D}$ in $\hat{\pi}$ is a set $\hat{\mathcal{D}}$ of lines in $\hat{\pi}$ through a point L with the property that for any two distinct lines a, b of $\hat{\pi}$ not through L for which the intersection of a and b belongs to a line of $\hat{\mathcal{D}}$ there is a Baer subplane $\hat{\pi}_0$ containing a, b, and L such that the lines of $\hat{\pi}_0$ through L are exactly the lines of $\hat{\mathcal{D}}$. Obviously $\hat{\pi}_0$ is uniquely determined by a, b, and $\hat{\mathcal{D}}$. We say that a Baer subplane of $\hat{\pi}$ belongs to $\hat{\mathcal{D}}$ if and only if it contains the lines of $\hat{\mathcal{D}}$. Let us denote by $\mathcal{B}$ the class of all Baer subplanes of $\hat{\pi}$ belonging to $\hat{\mathcal{D}}$. Take a line $g \in \hat{\mathcal{D}}$ and consider the affine plane $\mathcal{A}$ obtained from $\hat{\pi}$ by deleting g together with its points. Since any Baer subplane $\hat{\pi}_0$ of $\hat{\pi}$ belonging to $\hat{\mathcal{D}}$ contains g, the affine points and lines of $\mathcal{A}$ contained in $\hat{\pi}_0$ form an affine subplane $\mathcal{A}_0$ of $\mathcal{A}$ obtained from $\hat{\pi}_0$ by deleting g together with its points in $\hat{\pi}_0$. Denote by $\mathcal{C}$ the class of all affine Baer subplanes of $\mathcal{A}$ obtained in the above way from the subplanes of $\mathcal{B}$.

The affine points of any plane of $\mathcal{C}$ are points on the lines of $\hat{\mathcal{D}} - \{g\}$. Call $\mathcal{P}$ the set of all affine points of $\mathcal{A}$ on the lines of $\hat{\mathcal{D}} - \{g\}$. A triangle of $\mathcal{P}$ is a set of three non-collinear points in $\mathcal{P}$.

Our first step is to show:

(1) Any triangle of $\mathcal{P}$ is contained in exactly one plane of $\mathcal{C}$.

For, let A, B, C be any triangle of $\mathcal{P}$. At least two of the lines, say AB and AC, do not belong to $\hat{\mathcal{D}}$. The planes of $\mathcal{C}$ containing A, B, C are precisely the planes of $\mathcal{C}$ through AB and AC. However, from the definition of $\hat{\mathcal{D}}$ it follows that there is exactly one plane of $\mathcal{C}$ containing AB and AC.

(2) If two planes $\mathcal{A}_1$ and $\mathcal{A}_2$ of $\mathcal{C}$ have two affine points A, B in common, then they coincide in all their affine points on the line AB.

Here two cases can be distinguished:

(a) $AB \notin \hat{\mathcal{D}}$. According to the definition of $\hat{\mathcal{D}}$ the affine points of any plane of $\mathcal{C}$ through A and B are exactly the intersections of AB with the lines of $\hat{\mathcal{D}} - \{g\}$. Hence $\mathcal{A}_1$ and $\mathcal{A}_2$ have all their affine points in common.

(b) $AB \in \hat{\mathcal{D}}$. Suppose there is an affine point C of $\mathcal{A}_1 - \mathcal{A}_2$ on the line AB. Then AB is the only line of $\mathcal{A}_2$ through C. Consider the subplanes $\hat{\pi}_1$ and $\hat{\pi}_2$ of $\mathcal{B}$ from which $\mathcal{A}_1$ and $\mathcal{A}_2$ respectively were obtained. Since $\hat{\pi}_2$ is a Baer subplane, an arbitrary line h of $\hat{\pi}_1$ through C, distinct from AB, contains precisely one point H of $\hat{\pi}_2$. The point H belongs to a line $k \in \hat{\mathcal{D}}$ in $\hat{\pi}$. The lines h and k are both in $\hat{\pi}_1$; hence their intersection H is also a point of $\hat{\pi}_1$; furthermore H is different from B. Since there is exactly one subplane of $\mathcal{B}$ containing HA, HB, and $\hat{\mathcal{D}}$, the planes $\hat{\pi}_1$ and $\hat{\pi}_2$ coincide. Hence $\mathcal{A}_1$ and $\mathcal{A}_2$ must coincide. This contradiction shows that $\mathcal{A}_1$ and $\mathcal{A}_2$ have all their points on the line AB in common.

Call the set of all affine points of $\mathcal{A}$ common to all subplanes of $\mathcal{C}$ through two arbitrary distinct points A, B of $\mathcal{P}$ the segment $\overline{AB}$.

Denote by $\mathcal{T}$ the incidence structure whose points are the points of $\mathcal{P}$, blocks the segments determined by the pairs of distinct points from $\mathcal{P}$ with incidence defined as inclusion. Our aim is to prove that $\mathcal{T}$ is a three-dimensional affine space containing the planes of $\mathcal{C}$.

This will be done in several steps by the method applied in [6].

A linear manifold $\mathcal{M}$ of $\mathcal{T}$ is a subset of points in $\mathcal{T}$ such that with any two distinct points A, B of $\mathcal{M}$ all points of the block $\overline{AB}$ of $\mathcal{T}$ belong to $\mathcal{M}$. The linear manifold $\langle \mathcal{S} \rangle$ generated by a point set $\mathcal{S}$ of $\mathcal{T}$ is the intersection of all linear manifolds containing $\mathcal{S}$. Call the linear manifold of $\mathcal{T}$ generated by three arbitrary points, not on a common block, a plane of $\mathcal{T}$. Now we are able to show:

(3) <u>The points and the blocks of any plane of $\mathcal{T}$ form an affine plane.</u>

If the points A, B, C generating a plane $\pi(A, B, C)$ are not collinear in $\mathfrak{A}$ then there is exactly one plane $\mathfrak{A}_0$ of $\mathcal{C}$ containing them. It is easy to verify that $\mathfrak{A}_0$ is the linear manifold $\langle A, B, C \rangle$.

Assume that the points A, B, C generating a plane $\langle A, B, C \rangle$ are collinear in $\mathfrak{A}$. Since they are not contained in the same block of $\mathcal{T}$, this implies that the line h carrying A, B, C is in $\hat{\mathcal{D}} - \{g\}$. Let D be an affine point of $\mathfrak{A}$ on a line k of $\hat{\mathcal{D}} - \{g\}$, different from h. Denote by $\mathfrak{A}_1$ the plane of $\mathcal{C}$ through A, B, D. Take an arbitrary point X on g which is not an improper point of $\mathfrak{A}_1$. Since $\mathfrak{A}_1$ is a Baer subplane of $\mathfrak{A}$, the correspondence φ: Y $\to$ XY $\cap$ h where Y runs over all affine points of $\mathfrak{A}_1$ is a one-one mapping of the affine points of $\mathcal{A}_1$ onto the affine points of h. In order to verify that $\langle A, B, C\rangle$ is an affine plane whose lines are blocks of $\mathcal{T}$ it remains to show that the blocks of $\mathcal{T}$ on the line h are the images of the affine lines of $\mathcal{A}_1$ under φ. Let M, N be two arbitrary distinct points of h. Denote by U the unique point of XM in $\mathcal{A}_1$, and by V the unique point of XN in $\mathcal{A}_1$. If U = M and V = N then clearly $\overline{MN}$ is the image of the affine line UV of $\mathcal{A}_1$ under φ. Suppose $U \neq M$. Then U, V, M is a triangle, and according to (1) there is a unique plane $\mathcal{A}_2$ of φ containing it. Let $\hat{\pi}_2$ be the plane of $\mathcal{B}$ from which $\mathcal{A}_2$ was obtained. Together with M, U, V the plane $\hat{\pi}_2$ contains the points X and L, thus it contains the point $N = XV \cap ML$. Moreover the line UV of $\hat{\pi}_2$ is projected from X onto the line MN of $\hat{\pi}_2$. Thus the block $\overline{MN}$ is the image of the block $\overline{UV}$ under φ. Conversely, by the same argument every affine line of $\mathcal{A}_1$ is mapped under φ onto a block of $\mathcal{T}$ on h. This finishes the proof of (3).

$\mathcal{T}$ contains two types of planes: planes of type $\mathbb{B}$ which are the Baer subplanes of $\mathcal{C}$ and planes of type $\mathbb{L}$ which are the lines $\hat{\mathcal{D}} - \{g\}$. According to Lenz [12] the structure $\mathcal{T}$ is an affine space if the following properties are satisfied: (i) between the blocks an equivalence relation called parallelism is defined; (ii) for any point-block pair (P, b) there exists a unique block b' incident with P and parallel to b; (iii) for any four distinct points, A, B, C, D such that $\overline{AB}$ is parallel to $\overline{CD}$ and for any point $P \in \overline{AC}$, either $P \in \overline{CD}$ or $\overline{AB}$ and $\overline{PD}$ have a point in common. We shall define parallelism as follows:

($\mathbb{P}$) Two distinct blocks of $\mathcal{T}$ are <u>parallel</u> if and only if they are disjoint sets of points belonging to a common plane of $\mathcal{T}$. Each block is parallel to itself.

In view of (3) all planes of $\mathcal{T}$ are affine planes; hence properties (ii) and (iii) are satisfied. It remains to prove that parallelism ($\mathbb{P}$) is an equivalence relation. Clearly, ($\mathbb{P}$) is reflexive and symmetric; transitivity has to verified. Thus we have to show that:

(4) <u>If</u> b_1, b_2, <u>and</u> b_3 <u>are three blocks of</u> $\mathcal{T}$ <u>such that</u> b_1 <u>is parallel to</u> b_2 <u>and</u> b_2 <u>is parallel to</u> b_3, <u>then</u> b_1 <u>is parallel to</u> b_3.

For the proof of (4) the following cases can be distinguished:

CASE 1. b_1, b_2, b_3 are contained in a common plane of $\mathcal{T}$. Since the planes of $\mathcal{T}$ are affine planes, (4) follows immediately.

CASE 2. b_1 and b_2 are contained in a plane of type $\mathbb{B}$, and b_2 and b_3 are contained in a plane of type $\mathbb{L}$. Denote the plane through b_1 and b_2 by $\mathcal{A}_1$ and the plane through b_2 and b_3 by $\mathcal{L}_2$. Let ℓ be the line carrying the points of $\mathcal{L}_2$. Take two arbitrary points X, Y of b_1 and an arbitrary point Z of b_3. In view of (1) the points X, Y, Z determine a unique subplane $\mathcal{A}_3$ of $\mathcal{C}$. The plane $\mathcal{A}_3$ intersects ℓ in a block b_3' through Z having no point in common with b_2. In the affine plane $\mathcal{L}_2$ there is exactly one block through Z containing no point of b_2, namely the block b_3. Thus b_3 and b_3' coincide, that is b_1 and b_3 are parallel lines of $\mathcal{A}_3$. Hence b_1 and b_3 are parallel in $\mathcal{T}$.

CASE 3. b_1 and b_2 are contained in a plane $\mathcal{A}_1$ of type $\mathbb{B}$ and b_2 and b_3 also belong to a plane $\mathcal{A}_2$ of type $\mathbb{B}$. Denote the lines of $\hat{\mathcal{D}}$ carrying b_1, b_2, b_3 by ℓ_1, ℓ_2, ℓ_3 respectively. Since any line of $\hat{\mathcal{D}}$ contains lines of any subplane from $\mathcal{C}$, there is a line b_4 of $\mathcal{A}_1$ on ℓ_3. In view of (1) the blocks b_3 and b_4 are parallel in $\mathcal{T}$; obviously, b_4 is parallel to b_1. From the investigations in Case 2 it follows that b_1 and b_3 are parallel blocks of $\mathcal{T}$. This proves (4).

Thus $(\mathbb{P})$ is an equivalence relation, which implies:

(5) $\mathcal{T}$ is an affine space; the affine lines of $\mathcal{T}$ are the blocks of $\mathcal{T}$; the affine planes of $\mathcal{T}$ are Baer subplanes of $\mathcal{C}$ and the affine line of $\mathcal{A}$ belonging to $\hat{\mathcal{D}}$.

Consider an affine plane $\mathcal{L}_1$ of $\mathcal{T}$ represented by a line of $\hat{\mathcal{D}}$ - {g}. Take a point P of $\mathcal{T}$ in $\mathcal{A}_1$ and two distinct blocks

b_1, b_2 in $\mathcal{A}_1$ through P_1. For any other affine plane $\mathcal{L}_2$ of $\mathcal{T}$ represented by a line of $\hat{\mathcal{D}} - \{g\}$ and for any point P_2 of $\mathcal{L}_2$ there exist unique blocks b_3 and b_4 through P_2 parallel to b_1 and b_2 respectively. Hence the planes $\mathcal{L}_1$ and $\mathcal{L}_2$ are parallel in $\mathcal{T}$. On the other hand any affine plane of $\mathcal{T}$ represented by a Baer subplane of $\mathcal{B}$ intersects $\mathcal{L}_1$ in a block, that is in an affine line of $\mathcal{T}$. This implies that

(6) $\mathcal{T}$ is a three-dimensional affine space.

We have completed the proof of the following theorem:

THEOREM 3.1. *Let* $\hat{\pi}$ *be the dual plane of a derivable projective plane* π *and let* $\hat{\mathcal{D}}$ *be the dual of a derivation set* $\mathcal{D}$ *in* π. *If* $\mathcal{A}$ *is the affine plane obtained from* $\hat{\pi}$ *by deleting a line* $g \in \hat{\mathcal{D}}$ *together with its points, then the affine points of* $\mathcal{A}$ *on the lines of* $\hat{\mathcal{D}} - \{g\}$ *form a three-dimensional affine space* $\mathcal{T}$. *The affine lines of* $\mathcal{T}$ *are the affine lines of those Baer subplanes* $\mathcal{A}_i$ *of* $\mathcal{A}$ *which are obtained from the Baer subplanes of* $\hat{\pi}$ *belonging to* $\hat{\mathcal{D}}$ *after deleting* g *with its points. Any such subplane* $\mathcal{A}_i$ *is an affine plane of* $\mathcal{T}$.

As a corollary we obtain the following statement:

THEOREM 3.2. *If* $\hat{\pi}$ *is the dual plane of a derivable projective plane* π *and if* $\hat{\mathcal{D}}$ *is the dual of a derivation set*

$\mathcal{D}$ _of_ π _in_ $\hat{\pi}$, _then any Baer subplane of_ $\hat{\pi}$ _belonging to_ $\hat{\mathcal{D}}$ _is desarguesian_.

Our next aim is to consider Baer collineations of $\hat{\pi}$. A _Baer collineation_ of a projective plane is a collineation fixing pointwise a Baer subplane of the plane. Clearly, the Baer collineations fixing pointwise a given Baer subplane form a group.

Let $\hat{\pi}$, $\hat{\mathcal{D}}$, g, $\mathcal{A}$, $\mathcal{T}$, and $\mathcal{A}_i$ be defined as in Theorem 3.1. Denote by $\Pi(\mathcal{A}_i)$ the group of the Baer collineations fixing pointwise a Baer subplane $\mathcal{A}_i$ belonging to $\hat{\mathcal{D}}$. Any element α of $\Pi(\mathcal{A}_i)$ fixes the lines of $\mathcal{A}_i$; hence α fixes all lines of $\hat{\mathcal{D}}$. It follows that α induces a collineation in $\mathcal{T}$. Since $\mathcal{A}_i$ is a hyperplane of $\mathcal{T}$ the collineation $\bar{\alpha}$ induced by α in $\mathcal{T}$ is a perspectivity of $\mathcal{T}$ with axis $\mathcal{A}_i$. The collineation α fixes all lines of $\hat{\mathcal{D}}$; hence the centre of $\bar{\alpha}$ must lie on the improper line of $\mathcal{T}$ common to all affine planes $\mathcal{L}_i$ of $\mathcal{T}$ represented by the lines of $\hat{\mathcal{D}} - \{g\}$. In other words we have the following result:

THEOREM 3.3. _Let_ $\hat{\pi}$ _be the dual plane of a derivable projective plane_ π _and let_ $\hat{\pi}_0$ _be a Baer subplane of_ $\hat{\pi}$ _belonging to the dual of a derivation set of_ π _contained in_ $\hat{\pi}$. _If_ $\Pi(\hat{\pi}_0)$ _denotes the group of all Baer collineations of_ $\hat{\pi}$ _fixing_ $\hat{\pi}_0$ _pointwise, then_ $\Pi(\hat{\pi}_0)$ _is isomorphic to a subgroup of the group_ $\Gamma_{X,\mathcal{H}}$ _of all perspectivities in a three-dimensional projective space_ $\mathcal{S}$ _with a common axis_ $\mathcal{H}$

isomorphic to $\hat{\pi}_0$ *and with centres on a line* ℓ *of* $\mathcal{G}$ *not contained in* $\mathcal{H}$.

The investigations carried out for the duals $\hat{\pi}$ of derivable planes π can be dualized step by step. If G denotes an arbitrary but fixed point in a derivation set $\mathcal{D}$ of π, then it is not difficult to show that the lines of π through the points of $\mathcal{D}$ - {G} which are different from the line containing $\mathcal{D}$ are the points of a three-dimensional affine space $\hat{\mathcal{T}}$.

By completing $\hat{\mathcal{T}}$ to the corresponding projective space Σ it follows that the Baer subplanes of π belonging to $\mathcal{D}$ are hyperplanes of Σ. Thus the statements of Theorems 3.2 and 3.3 can be formulated for derivable planes:

THEOREM 3.4. *Any Baer subplane belonging to a derivation set of a derivable projective plane is desarguesian.*

THEOREM 3.5. *Let* $\Pi(\pi_0)$ *be the group of the Baer collineations of a derivable projective plane* π *fixing pointwise a Baer subplane* π_0 *which belongs to a derivation set of* π. *Then* $\Pi(\pi_0)$ *is isomorphic to a subgroup of the group* $\Gamma_{X,\mathcal{H}}$ *of the perspectivities in a three-dimensional projective space* Σ *with a common axis* $\mathcal{H}$ *isomorphic to* π_0 *and with centres* X *on a line* ℓ *of* Σ *not contained in* $\mathcal{H}$.

REFERENCES

1. S.A. Amitsur. Rational identities and applications to algebra and geometry. J. Algebra 3 (1966): 304-359.
2. J. André. Über nicht-desarguessche Ebenen mit transitiver Translationsgruppe. Math. Z. 60 (1954): 156-186.
3. R. Baer. Homogeneity of projective planes. Amer. J. Math. 64 (1942): 137-152.
4. R.H. Bruck and R.C. Bose. The construction of translation planes from projective spaces. J. Algebra 1 (1964): 85-102.
5. —— Linear representations of projective planes in projective spaces. J. Algebra 4 (1966): 117-172.
6. J. Cofman. Baer subplanes of affine and projective planes. Math. Z. 126 (1972): 339-344.
7. P. Dembowski. Finite Geometries. New York: Springer (1968).
8. M. Hall. Projective planes. Trans. Amer. Math. Soc. 54 (1943): 229-277.
9. A. Herzer. Dualitäten mit zwei Geraden aus absoluten Punkten in projektiven Ebenen. Math. Z. 129 (1972): 235-257.
10. —— Ableitung Zweier selbstdualer, zum Satz von Pappos äquivalenter Schließungssätze aus der Konstruktion von Dualitäten. Geom. Dedicata 2 (1973): 283-310.
11. —— Charakterisierung verschiedener regulärer Faserungen durch Schließungssätze. To appear in Archiv der Mathematik.
12. H. Lenz. Zur Begründung der analytischen Geometrie. Sitz-Ber. Bayer. Akad. Wiss. (1954): 17-72.
13. —— Vorlesungen über projektive Geometrie. Akad. Verl. Ges. Geest und Portig, Leipzig (1965).
14. T.G. Ostrom. Semi-translation planes. Trans. Amer. Math. Soc. 111 (1964): 1-18.
15. G. Pickert. Projektive Ebenen. Berlin-Göttingen-Heidelberg: Springer (1955).
16. O. Prohaska. Endliche ableitbare affine Ebenen. Geom. Dedicata 1 (1972): 6-17.

UNITARY PLANES

Martin Götzky

In the following the intention is to develop unitary geometry of the plane, with the help of quasisymmetries. At the heart of the considerations is the theorem of the third quasireflection (see Götzky, 1973). Part I deals with unitary-euclidean geometry, and part II with unitary-minkowskian geometry (see Götzky, 1970).

I. GENERALIZED EUCLIDEAN GEOMETRY

Let P be a projective plane and S a set of perspective collineations $\neq 1$. If a is a straight line of P, let S(a) be the set of elements of S with axis a. Assume the following property:

(*1) $\{1\} \cup S(a)$ is a group for every straight line a.

Obviously $\{S(a);$ a is a line of $P\}$ is a partition of S.

Let G = $\langle S \rangle$ be the group generated by S. We are interested in generated groups (G, S) with (*1), for which the following statement holds:

THEOREM OF THE THIRD QUASIREFLECTION. Let a, b, c be straight lines through a given point O with $a \neq b$. Let $\alpha \in S(a)$, $\beta \in S(b)$, and $c\alpha\beta \neq c$. Then there exists a $\gamma \in S(c)$ and a straight line d passing through O such that $\alpha\beta\gamma \in S(d)$.

Fixing a straight line g of P we may consider the affine plane $A = P_g$ instead of P. If S_0 is the set of line-reflections of A and A a pappian plane with Fano axiom, then Veblen (Veblen and Young, 1918, vol. 2, §52) has already stated this theorem for $(\langle S_0\rangle, S_0)$. Only (*1) does not hold.

There are many examples of generated groups (G, S) with (*1) for which the theorem of the third quasireflection holds. For instance let S_2 be the set of all axial collineations of $\langle S_1\rangle$ with S_1 the set of all shears and line-reflections of an affine Moufang plane $A = P_g$ with Fano axiom. Then (G, S_i) for i = 1, 2 is a generated group with (*1) holding. The theorem of the third quasireflection holds for (G, S_2) exactly if A is desarguesian and for (G, S_1) exactly if A is pappian (Götzky, 1972).

For the case (G, S_2), let $\alpha \in S(a)$, $\beta \in S(b)$, and $\gamma \in S(c)$ with a, b, c and O, P, Q as shown in figure 1. Further let $Q = Q^{\alpha\beta\gamma}$. Then $P = P^{\alpha\beta\gamma}$ for every point P on the connecting line P of O and Q if and only if $\alpha\beta\gamma \in S$.

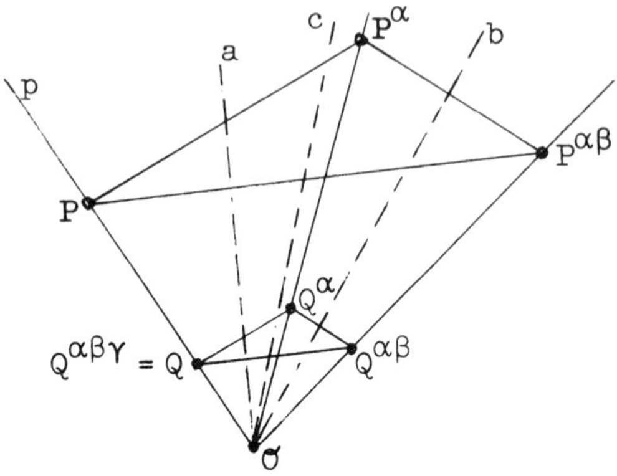

FIGURE 1

This shows that quasireflection is closely related to Desargues' theorem.

Now let $V = V_n(K, f, \iota)$ be an n-dimensional left vector space where K is a skew field of char. $\neq 2$, ι is an anti-automorphism of K with $\iota^2 = 1$, and f is an ι-hermitian form, that is a function mapping each pair a, b of vectors onto an element of K and satisfying

$$f(sa, b) = s\, f(a, b) \quad \text{if} \quad s \in K,$$

$$f(a' + a'', b) = f(a', b) + f(a'', b),$$

$$f(a, b)\iota = f(b, a).$$

(If $\iota = 1$, then K is commutative and f a bilinear form, hence V is a metric (orthogonal) vector space.)

The set of all linear mappings π satisfying $f(a^{\pi}, b^{\pi}) = f(a, b)$ is a group called the unitary group U(V). The subset of all linear mappings π of U(V) fixing each element of rad V := $\{x \in V;\ f(x, V) = 0\}$ is also a group called the restricted unitary group U*(V).

Let $a \notin \text{rad } V$ and $s \in K \setminus \{0\}$, then

(1) $x\sigma(s, a) := x - f(x, a)\cdot s\cdot a$

defines a linear map $\sigma(s, a)$ fixing the hyperplane $\{y;\ f(y, a) = 0\}$ vectorwise. Also, $\sigma(s, a) \in U^*(V)$ if and only if

(2) $s^{-1}\iota + s^{-1} = f(a, a)$.

Let $S(Ka) := \{\sigma(s, a);\ s^{-1}\iota + s^{-1} = f(a, a)\}$ and

$$S(V) = \bigcup_{a \notin \text{rad } V} S(Ka).$$

Obviously $S(Ka) \cup \{1\}$ is a group.

THEOREM. Consider the generated restricted unitary group $(U^*(V), S(V))$. Then, for the one-dimensional subspaces of V not contained in rad V, we have: if $Ka \neq Kb$ and $Kc \leq Ka + Kb$ and also $\alpha \in S(Ka)$, $\beta \in S(Kb)$, and $(Kc)\alpha\beta \neq Kc$, then there exist a $\gamma \in S(Kc)$ and a subspace $Kd \leq Ka + Kb$ with $\alpha\beta\gamma \in S(Kd)$ (Götzky, 1964, 1965).

Let $n = 3$ and $P = PV$ be the desarguesian projective plane with one-dimensional subspaces of V as straight lines, two-dimensional subspaces as points, and incidence defined by inclusion. Then, for P, $(PU^*(V), PS(V))$ is a generated group satisfying (*1) and by the theorem also satisfying the theorem of the third reflection.

REMARK. If S contains only reflections, the theorem of the third quasireflection for (<S>, S) becomes the theorem of the three reflections (see Bachmann, 1959 or 1973).

$P = \mathrm{PV}$ is called a unitary plane. By $Kx \perp Ky$ if $f(x, y) = 0$, an orthogonality relation for these planes is defined. If dim rad V = 1, $P_{\mathrm{rad}\ V}$ is a unitary affine plane. In this case we say

$P_{\mathrm{rad}\ V}$ is unitary-euclidean if index f = 0,

$P_{\mathrm{rad}\ V}$ is unitary-minkowskian if index f = 1.

REMARK. If index f = 0 for (PU*(V), PS(V)), the theorem of the third quasireflection holds even if we delete the assumption $c\alpha\beta \neq c$; we say the stronger theorem of the third reflection holds. Also the theorem of the third quasireflection keeps holding if $c\alpha\beta \neq c$ is replaced by $c \not\perp c$. In this case we say that the varied theorem of the third qusireflection holds.

Next we concentrate on affine translation planes $A = P_g$, with a point involution j on g which defines an affine orthogonality relation by $a \perp b$ if the intersecting points of a with g and b with g correspond under j (see figure 2).

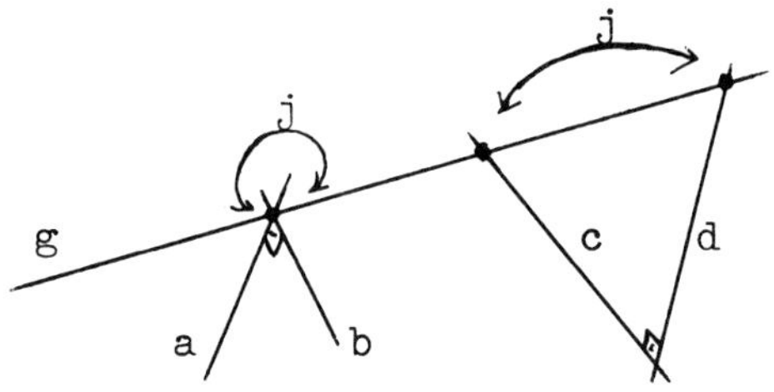

FIGURE 2

To avoid distinction between cases we restrict ourselves to cases with j free of fixed points. Thus there are no self-orthogonal straight lines.

Assume a system S with (*1) such that:

(*2) S(a) contains an involution for all straight lines a.

Obviously then there is a unique reflection on every straight line a of P_g and Char $P_g \neq 2$; that is, the Fano axiom holds.

We assume that the stronger theorem of the third quasireflection holds for (G, S) with G = <S> and assume furthermore that

(*3) S contains all perspective collineations of G.

(*4) G preserves the orthogonality induced by j.

The question is whether A is unitary or not (if unitary it is unitary-euclidean). For a better understanding of this question, I shall prove some little statements. (For the following discussions if β, γ are elements of the same group, we define $\beta^{\gamma} := \gamma^{-1}\beta\gamma$.)

LEMMA 1. The point reflections are in G.

LEMMA 2. Suppose $\alpha \in S(a)$, $\beta \in S(b)$, and a $\|$ b. Then $\alpha\beta$ is a translation if and only if there is a point reflection C such that $\alpha\beta^{C} = 1$.

PROOF. $\alpha^{-1}\alpha^{C} = C^{\alpha}C$ is a translation. Let $\alpha\beta$ be a translation and C a point reflection with $b^{C} = a$. Then

$$\alpha\beta^{C} \cdot (\beta^{-1})^{C} = \alpha\beta$$

where $(\beta^{-1})^{C}\beta$ and $\alpha\beta$ are translations, so $\alpha\beta^{C}$ is a translation. Moreover, $\alpha\beta^{C} \in S(a)$, so $\alpha\beta^{C} = 1$.

LEMMA 3. <u>Let</u> $\alpha \in S(a)$, $\beta \in S(b)$ <u>with</u> $O|\alpha \neq \beta|O$. <u>Then there is a</u> $\delta \in S(d)$ <u>for all</u> d <u>such that</u> $\delta\alpha\beta$ <u>is a glide-quasireflection</u>.

REMARK. If $\alpha \in S \cup \{1\}$ then α is called a <u>quasireflection</u>. If τ is a translation and $\alpha \in S \cup \{1\}$, then $\tau\alpha$ is called a <u>glide-quasireflection</u>.

PROOF. Let C be a point reflection such that d^{C} I O. There is a $\delta \in S(d)$ such that $\gamma := \delta^{C}\alpha\beta \in S$. If $\gamma \in S(c)$ then c I O (theorem of the third quasireflection). Now $\delta\alpha\beta = \delta(\delta^{-1})^{C}\gamma$ is a glide-quasireflection.

DEFINITION 4. If $\pi \in G$ and M_{XY} is the midpoint of the points XY, then $H_{\pi}: X \to M_{XY}$ defines a point map called the <u>midpoint map</u> of π.

LEMMA 5. <u>If a midpoint map</u> H_{π} <u>preserves collinearity, then it is a collineation or else there is a line passing through all the points of image</u> H_{π}.

PROOF. See Götzky (1972).

LEMMA 6. H_α __preserves collinearity for all__ $\alpha \in S$.

From now on, we shall not distinguish between a point and its point reflection.

PROOF OF LEMMA 6. Let $\alpha \in S(a)$, $O \text{ I } a$, and $O \text{ I } c \text{ I } P$ but $c \neq a$. We show that, regardless of how P was chosen, there is a line m passing through O such that P^{H_α} is on m.

Consider the identity $O \cdot POP^\alpha = (PP^{\alpha O})^O = P^{\alpha O}P$. Let $b \perp a$ and $b \text{ I } O$ such that O is the product of the line reflections in a and b. Then $\alpha O = \alpha'\beta$ for some $\alpha' \in S(a)$ and $\beta \in S(b)$. By the theorem of the third quasireflection, there is a $\gamma \in S(c)$ such that $\gamma := \gamma\alpha'\beta \in S$. (See figure 3.) Obviously $P^{\alpha O} = P^\delta$, so $O \cdot POP^\alpha = P^\delta P$. Let $\delta \in S(d)$. Then d is a line through O perpendicular to the connecting line of O and POP^α. Since d does not depend on the choice of P on c, the connecting line also does not. If

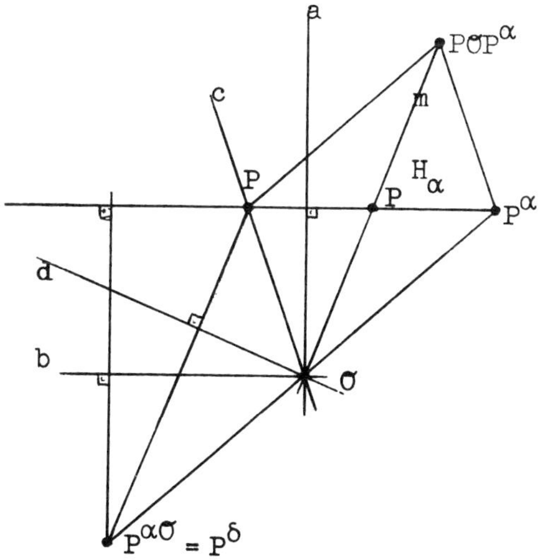

FIGURE 3

this connecting line is called m, the lemma is established.

LEMMA 7. *If* $\omega \in G$ *and* τ *is a translation, then* $H_{\omega\tau} = H_\omega H_\tau$.

PROOF. Using $\tau = H_\tau H_\tau$ we get

$$P^{M_{PP^{\omega\tau}}} = P^{\omega\tau} = P^{M_{PP^{\omega\cdot\tau}}} = P^{H_\tau^{-1} M_{PP^\omega} H_\tau}.$$

The uniqueness of the midpoint yields

$$\begin{array}{ccc} M_{PP^{\omega\tau}} & = & (M_{PP^\omega})^{H_\tau} \\ \| & & \| \\ P^{H\omega\tau} & & (P^{H_\omega})^{H_\tau} = P^{H_\omega H_\tau} \end{array}$$

This proves the lemma.

LEMMA 8. *If* τ *is a translation and if* $\omega \in S$, *then* $H_{\tau\omega}$ *and* $H_{\omega\tau}$ *preserve collinearity*.

PROOF. Since S contains all the axial transformations by (*3), it follows that $\tau\omega\tau^{-1} \in S$. Furthermore, we have $H_{\tau\omega} = H_{\tau\omega\tau^{-1}\cdot\tau}$. Hence we only have to prove that $H_{\omega\tau}$ preserves collinearity. But $H_{\omega\tau} = H_\omega H_\tau$ and H_ω, H_τ both preserve collinearity.

LEMMA 9. *If* $\alpha, \beta \in S$ *then* $H_{\alpha\beta}$ *preserves collinearity*.

PROOF. Let $\alpha \in S(a)$ and $\beta \in S(b)$. If $a \| b$ then $\alpha\beta \in S \cup \{\tau;\ \tau \text{ translation}\}$, hence $\alpha\beta$ preserves collinearity. Thus we may assume that

$$a \neq b \text{ and } a, b \text{ I } O \text{ for some point } O.$$

Let c be a line. There is a $\gamma \in S(c)$ such that $\gamma\alpha\beta$ is a glide-reflection (Lemma 3). Now $P^{\alpha\beta} = P^{\gamma\alpha\beta}$ and hence $P^{H_{\alpha\beta}} = P^{H_{\gamma\alpha\beta}}$ for all P I c.

Since $H_{\gamma\alpha\beta}$ preserves collinearity, then $H_{\alpha\beta}$ also has to preserve collinearity.

LEMMA 10. <u>Let O, P, Q be non-collinear points</u>. <u>There exists an axial collineation</u> δ <u>such that</u> $O^{\delta} = O$ <u>and</u> $P^{\delta} = Q$.

PROOF. Let O, P, Q be non-collinear and a, b, c, R be as shown in figure 4.

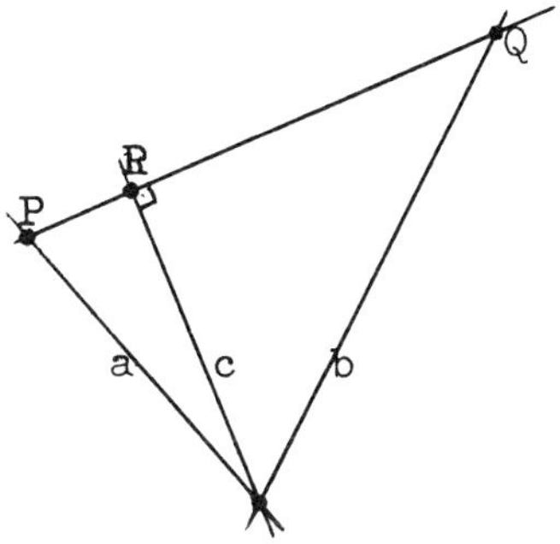

FIGURE 4

(1) Assume $|S(x)| > 1$ for all lines x.

Choose $\alpha' \in S(a)$, $\beta'' \in S(b)$, and $\gamma \in S(c)$ such that γ is an involution but α' and β' are not.

Since $\alpha'^{\gamma} := \gamma^{-1}\alpha'\gamma$ is a non-involution, Lemmas 5 and 6 together yield that $H_{\alpha'^{\gamma}}$ is a collineation with axis a^{γ}. It is easy to show that Lemmas 5 and 9 together yield that $H_{\alpha'\gamma}$ is a collination (use the fact that a $\not I$ C).

Now $X^{\alpha'\gamma} = X^{\alpha'\gamma}$ for all X I c, so $X^{H_{\alpha'\gamma}} = X^{H_{\alpha'\gamma}}$ for all X I c. Therefore $\gamma' := H_{\alpha'\gamma}H^{-1}_{\alpha'\gamma}$ fixes every point of c; that is, γ' is axial with axis c (it might not be in S(c)). Similarly, $\gamma'' := H_{\beta''\gamma}H^{-1}_{\beta''\gamma}$ is an axial collineation with axis C. Because $H_{\alpha'\gamma} = H^{\gamma}_{\alpha'} = \gamma^{-1}H_{\alpha'}\gamma$ and $\gamma^{-1} = \gamma$, we obtain

$$R = P^{H_{\alpha'\gamma}} = P^{\gamma'\gamma H_{\alpha'}\gamma} \underset{P\ I\ a}{=\!=\!=} P^{(\gamma'\gamma)^{H_{\alpha'}}\cdot\gamma},$$

$$R = Q^{H_{\beta''\gamma}} = Q^{\gamma''\gamma H_{\beta''}\gamma} \underset{Q\ I\ b}{=\!=\!=} Q^{(\gamma''\gamma)^{H_{\beta''}}\cdot\gamma}.$$

Using R I c, we are led to

$$R = P^{(\gamma'\gamma)^{H_{\alpha'}}} = Q^{(\gamma''\gamma)^{H_{\beta''}}}.$$

Since $(\gamma'\gamma)^{H_{\alpha'}}$ and $(\gamma''\gamma)^{H_{\beta''}}$ are obviously axial and both have the same direction PQ, the product

$$\delta := (\gamma'\gamma)^{H_{\alpha'}}\left[(\gamma''\gamma)^{H_{\beta''}}\right]^{-1}$$

is also axial. This δ satisfies $O^{\delta} = O$ and $P^{\delta} = Q$, as required.

(2) Assume $|S(a)| = 1$ for one a I O. Let b, c I O, $a \neq b \neq c \neq a$. Suppose, if possible, that $|S(c)| > 1$ and choose $\alpha \in S(a)$, $\beta \in S(b)$, and distinct $\gamma, \gamma'' \in S(c)$.

The theorem of the third quasireflection yields that $\delta = \gamma\beta\alpha$ and $\delta' = \gamma'\beta\alpha$ are both in S. But this would lead to the sequence of inferences $\delta'\delta^{-1} = \delta'\delta^{-1} \in S(c) \cup \{1\}$; $\gamma, \gamma', \delta, \delta' \in S(c)$; hence $\beta\alpha \in S(c)$, which is impossible.

So the assumption $|S(c)| > 1$ was incorrect. This proves that $|S(c)| = 1$ for all lines c. But in this case the theorem of the third quasireflection yields the theorem of the three reflections. Hence $\mathcal{A}$ is pappian (see Bachmann, 1959 or 1973) and Lemma 10 holds.

THEOREM. $\mathcal{P}$ <u>is a Moufang plane</u>.

PROOF. It is sufficient to show that for every pair $a \nparallel b$ in $\mathcal{A} = \mathcal{P}_g$ there is a line reflection with axis a in the direction of b (Pickert, 1955). Since there are line reflections in $\mathcal{A}$ it is therefore sufficient to show that $Aut(\mathcal{A})$ is transitive on the non-parallel pairs of lines of $\mathcal{A}$.

Thus we want to map $a \nparallel b$ onto $c \nparallel d$. Since it is a translation plane, we may assume that a, b, c, d I O for a point O. Using Lemma 10 we furthermore may assume that a = c.

Fix e, P, Q as in figure 5 ($c \parallel d$). By Lemma 10, there is an m and a $\mu \in S(m)$ such that $O^{\mu} = \sigma$ and $P^{\mu} = Q$; hence $a^{\mu} = a = c$ and $b^{\mu} = d$. This proves the theorem.

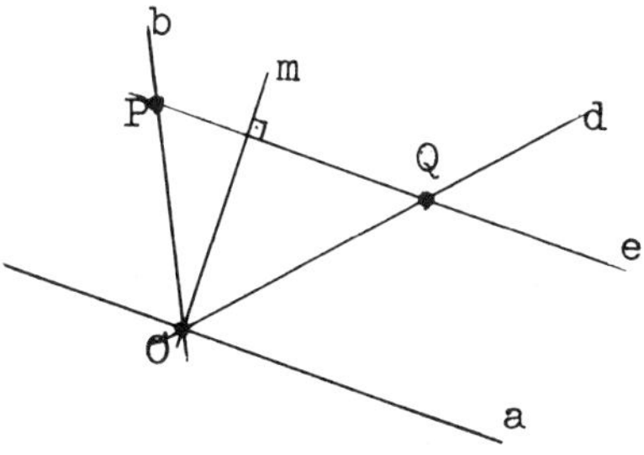

FIGURE 5

Whether the plane A is unitary under the given conditions now depends on the existence of both a non-desarguesian affine Moufang plane $A = P_g$ and a given point-involution j on g that is free of fixed points, which together admit a generated collineation group (G, S) satisfying (*1) to (*4) and the theorem of the third quasireflection. It is still unknown whether such planes exist.

In the next section, unitary-minkowskian planes $A = P_g$ are investigated. Using an additional condition denoted there as A.6, a full characterization of those planes is given.

II. UNITARY-MINKOWSKIAN PLANES

II.1. Axioms

The purpose of this section is to introduce the group plane of a given generated group (G, S) with S the set of generators of G, and further to present a set of axioms for the elements of S which ensures that the group plane has certain desired properties.

BASIC ASSUMPTION. Let (G, S) be a generated group with $1 \notin S$ such that the following properties hold:

(1) There is a partition $S_G = \{S(x);\ x \in G\}$ of S which is invariant under inner automorphisms.

(2) $S(x) \cup \{1\}$ is a group for all $x \in G$.

(3) For all $x \in G$ such that $S(x) \cup \{1\}$ is non-abelian, the set $S(x)$ contains an involution.

(4) If $a \neq b$ and $\beta \in S(b)$, then $S(a)^{\beta} = S(b)$ yields $S(b) \subseteq C_G(S(a) \cup \{1\})$.

(Remember that $S(a)^{\beta} := \beta^{-1}S(a)\beta$.)

DEFINITION 1.1. $\mathcal{P} := \{\alpha\beta;\ \alpha, \beta \in S$ and $\alpha, \beta, \alpha\beta$ involutory$\}$.

Suppose $S(a)$ and $S(b)$ are elements of the partition $S_{\mathcal{G}}$ and $A \in \mathcal{P}$; and consider the following properties:

(a) $A \in C_G(S(b) \cup \{1\})$.

(b) $A \notin S(b)$.

(c) $S(a) \cup \{1\} \leq C_G(S(b) \cup \{1\})$.

(d) If $S(a)$ or $S(b)$ contains an involution, then $a \neq b$.

DEFINITION 1.2. $A \mid b$ precisely when (a) and (b) hold. $a \mid b$ precisely when (c) and (d) hold.

The triplet $(\mathcal{P}, \mathcal{G}, \mid)$ with $\mathcal{P}$ a point set, $\mathcal{G}$ a line set and $\mid$ as both an incidence relation and an orthogonality relation is now an incidence structure with orthogonality relation between lines. This incidence structure is called the <u>group plane</u> $E(G, S_{\mathcal{G}})$.

From now on let (G, S) always satisfy the basic assumption. We are interested in the following set of axioms:

A.1. If $A, B \in P$, there is a $c \in G$ such that $c \mid A, B$.

A.2. If $a, b \in G$, either there is a $c \in G$ such that $c \mid a, b$ or there is a $C \in P$ such that $C \mid a, b$.

A.3. If $A, B \in P$ and $a, b \in G$ and $A, B \mid a, b$, then $A = B$ or $a = b$.

A.4. Let $a, b, c \in G$ and $O \in P$ such that $a \neq b$ and $O \mid a, b, c$. Further let $\alpha \in S(a)$ and $\beta \in S(b)$. If $S(c)$ contains an involution, then there exist a $\gamma \in S(c)$ and a $d \mid O$ such that $\delta := \alpha\beta\gamma \in S(d)$.

A.5. Let $\alpha \in S(a)$, $\beta \in S(b)$, and $a, b \mid d$ where $a, b, d \in G$. Then one and only one of the following conditions hold:

(a) $\alpha\beta \in S$.

(b) There is a $C \in P$ such that $\alpha\beta^C = 1$.

A.6. Let $a, b, c \in G$ and $D \in P$ such that $a, b, c \mid D$. If neither $S(a)$ nor $S(c)$ contains an involution and if further $a \neq b \neq c$, then there is a $\beta \in S(b) \cup \{1\}$ such that $S(a)^\beta = S(c)$.

Ex. A. There exist $A \in P$ and $b \in G$ such that $b \mid b$ but $A \mid b$.

($A \nmid b$ means that $A \mid b$ does not hold. $a \nmid b$ means that $a \mid b$ does not hold.)

THEOREM 1.3. <u>The group plane</u> $E = E(G, S_G) = (P, G, \mid)$ <u>is a Fano translation-plane</u> (<u>that is, a translation plane satisfying the Fano axiom</u>) <u>with an orthogonality relation definable by a point involution</u> j <u>on the line of infinity</u>.

The involution j *has at least four fixed points.*

For all $\omega \in G$, *let* $\bar{\omega}$ *be the inner automorphism belonging to* ω. *Then* $(\bar{G}, \bar{S})$ *is a generated group, with* $\bar{S}$ *a set of axial collineations of* E, *and the following properties are satisfied:*

(*1) $\bar{S}(a) \cup \{1\}$ *is a group for every* $a \in G$.

(*3) *Each axial collineation of* $\bar{G}$ *is in* $\bar{S}$.

(*4) $\bar{G}$ *preserves the orthogonality of* E.

(*5) *If* $a \in G$ *satisfies* $a \mid a$, *then* $S(a) \cup \{1\}$ *is a group of shears with axis* a. *If* $a \in G$ *satisfies* $a \nmid a$, *then* $S(a) \cup \{1\}$ *is a group of homologies with axis* a *containing an involution.*

Theorem 1.3 holds for generated groups (G, S) satisfying the basic assumption, A.1 to A.6, and Ex. A (see Götzky, 1970). The proof needs a large number of elementary properties of the partition S_G, which follow from the assumptions. It includes a proof of the existence and uniqueness of the perpendicular, and also a proof of the fact that a quadrilateral with three right angles is a rectangle.

From now on, let (G, S) always satisfy the basic assumption along with A.1 to A.6 and Ex. A so that, in particular, Theorem 1.3 holds.

REMARKS. Axiom A.4 yields the varied theorem of the third quasireflection for the plane E. By A.6, the group

$\overline{S(a)} \cup \{1\}$ acts transitively on the set of self-orthogonal lines different from a which pass through a given point O on a.

II.2. Reduction theorem

Each element of G is a product of elements of S. We are interested in representations of a given element $\omega \in G$ as a product of elements of S with a minimum of factors.

Suppose that a, b are non-parallel lines of the group plane E and c is a further line satisfying $c \not\perp c$. Let $\alpha \in S(a)$, $\beta \in S(b)$ and let O be the point with $O \mid a, b$. Finally let $d \| c$ with $d \mid O$. By A.4, there are a $\delta \in S(d)$ and a line $m \mid O$ such that $\alpha\beta = \mu\delta$ for some $\mu \in S(m)$.

Using this result, one easily proves:

LEMMA 2.1. *Let* $\omega \in G$. *There are lines* a, b, c, d *satisfying* $a \| b$, $c \| d$, *and* $d \not\perp d$, *such that for a certain choice of* $\alpha \in S(a)$, $\beta \in S(b)$, $\gamma \in S(c)$, *and* $\delta \in S(d)$ *the following holds*:

$$\omega = \alpha\beta\gamma\delta.$$

Using A.5, one deduces from Lemma 2.1 the following:

LEMMA 2.2. *Let* $\omega \in G$. *Then either there are non-parallel lines* a, b *such that, for certain* $\alpha \in S(a)$ *and* $\beta \in S(b)$, *we have* $\omega = \alpha\beta$; *or there are two points* A, B *and a line* c *such that, for a certain* $\gamma \in S(c)$, *we have* $\omega = AB\gamma$.

(If $\bar{\alpha}$, $\bar{\beta}$ are axial collineations such that their axes have a point in common, then $\bar{\alpha}\bar{\beta}$ is called a *quasirotation*. The elements $\bar{A}\bar{B}\bar{\gamma} \in \bar{P}\bar{P}\bar{S}$ are called *glide-quasireflections* if $\bar{\gamma}$ is a quasireflection, or *glide-shears* if $\bar{\gamma}$ is a shear.)

Lemma 2.2 tells us that the elements $\bar{\omega}$ of $\bar{G}$ are either quasirotations or glide-quasireflections, or glide-quasishears. Because $\bar{\omega}$ is also the inner automorphism of G belonging to ω, Lemma 2.2 yields:

THEOREM 2.3. *The centre of* G *is trivial and therefore* $\bar{G} \simeq G/_{Z(G)} = G$.

Lemma 2.2 also yields

REDUCTION THEOREM 2.4. *If* $O \in P$ *is any point of* E, *then* $G = C_G(O)PP$.

II.3. The theorem of the antiorthological i-quadrilaterals

The theorem referred to in the title of this section is a special case of the theorem of the antiorthological quadrilaterals, due to Schütte (1955), which characterizes the affine orthogonality relations induced by hermitian forms.

DEFINITION 3.1. Let i = 1, 2, 3, 4 and j = 1, 2. Consider 4 lines a_i and 2 self-orthogonal lines s_j such that there are points A_1, A_2, A_3, A_4 satisfying both: $a_i, a_{i-1} \mid A_i$

with $a_0 := a_4$, and $s_j \mid A_j, A_{j+2}$. Then $(a_1, a_2, a_3, a_4; s_1, s_2)$, which may be abbreviated as $(a_i; s_j)$, is called an i-<u>quadrilateral</u>. The a_i are the <u>sides</u> and the s_j the diagonals of the quadrilateral. If there is a point $A \mid s_1, s_2$ then (a_i, s_j) is called almost complete and A is referred to as the <u>diagonal point</u> of the i-quadrilateral (see figure 6).

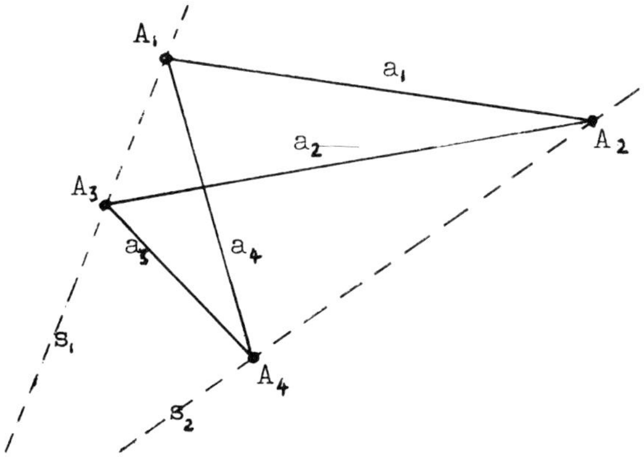

FIGURE 6

The following two lemmas are the core of the theory to be considered here:

LEMMA 3.2 (first lemma of the i-quadrilaterals). <u>Let</u> $(\alpha_i; s_j)$ <u>be a non-degenerate</u>, <u>almost complete</u> i-<u>quadrilateral with vertices</u> A_i <u>and diagonal point</u> A. <u>For</u> i = 1, 2, 3 <u>let</u> b_i <u>be the line satisfying</u> $b_i \mid A, a_i$. <u>Let</u> b_4 <u>be a line through</u> A. <u>Then the following two statements are equivalent</u>:

(1) $b_4 \mid a_4$.

(2) _If_ $\beta_i \in S(b_i)$ _satisfies_ $s_1^{\beta_i} = s_2$ _for_ $i = 1, 2, 3, 4$, _then_ $\beta_1\beta_2^{-1}\beta_3\beta_4^{-1} \in S(s_1)$.

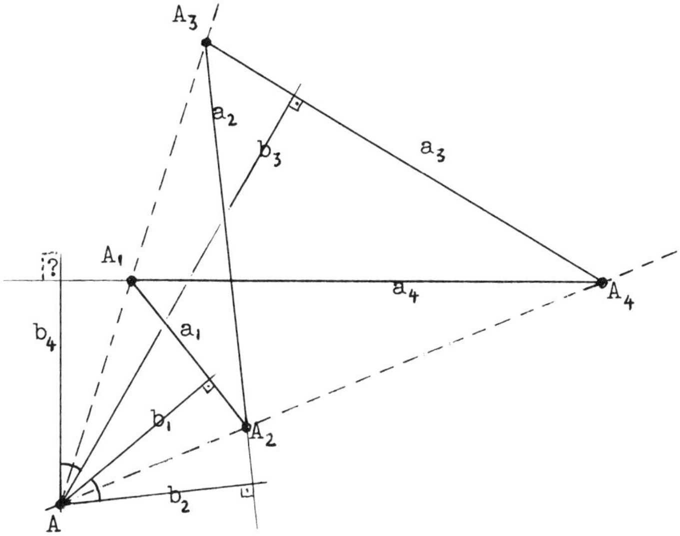

FIGURE 7

PROOF. There exists exactly one line satisfying (1). Assume if possible that there are two elements $\beta_4, \beta_4' \in S$ such that (2) holds. Then $\beta_4'\beta_4^{-1} \in S(s_1)$ and, since $s_1^{\beta_4'} = s_1^{\beta_4} = s_2$ implies $b_4 \neq s_1 \neq b_4'$, it follows that $\beta_4'\beta_4^{-1} = 1$. So there is at most one $\beta_4 \in S$ satisfying (2), which means that it is sufficient to show that (1) implies (2).

Assume $\beta_1, \beta_2, \beta_3, \beta_4$ satisfy (1) and the assumptions of (2). Let $\omega = \beta_1\beta_2^{-1}\beta_3\beta_4^{-1}$. Then for $i = 1, 3$ we have the relations:

$$A_{i+1} \mid a_i, s_2 \quad \text{and} \quad A_i \mid a_i, s_1.$$

It follows that:

(a) $A_{i+1}, A_i^{\beta_i} \mid a_i, s_2$ for $i = 1, 3$

and that

(b) $A_{i+1}, A_i^{\beta_i^{-1}} \mid a_i, s_1$ for $i = 2, 4$ and $A_5 = A_1$.

Now (a) yields $A_{i+1} = A_i^{\beta_i}$ for $i = 1, 3$ and (b) yields $A_{i+1} = A_i^{\beta_i^{-1}}$ for $i = 2, 4$. So $A_1^{\omega} = A_1$.

Reviewing the situation we know that $A^{\omega} = A \neq A_1 = A_1^{\omega}$ and $s_1 \mid A, A_1$. Thus Lemma 2.2 yields $\omega \in S(s_1)$. This proves the lemma.

LEMMA 3.3 (second lemma of the i-quadrilaterals). _Let_ (a_i, s_j) _be a non-degenerate, almost complete i-quadrilateral with vertices_ A_i _and diagonal point_ A. _For_ $i = 1, 4, 3$ _let_ b_i _be the line satisfying_ $a_i \parallel b_i \mid A$. _Let_ b_2 _be any line passing through_ A. _Then the following statements are equivalent:_

(1) $b_2 \parallel a_2$.

(2) _If_ $\beta_i \in S(b_i)$ _satisfies_ $s_1^{\beta_i} = s_2$ _for_ $i = 1, 2, 3, 4$, _then_ $\beta_1\beta_4^{-1}\beta_3\beta_2^{-1} \in S(s_1)$.

PROOF. As in the proof of 3.2 it is sufficient to show that (1) implies (2).

Assume $\beta_1, \beta_2, \beta_3, \beta_4$ satisfy (1) and the assumptions of (2). Choose $\alpha_i \in S(a_i)$ such that $s_1^{\alpha_i} \parallel s_2$. Define $\nu = \alpha_1\alpha_4^{-1}\alpha_3\alpha_2^{-1}$ and $\omega = \beta_1\beta_4^{-1}\beta_3\beta_2^{-1}$. We prove a sequence of statements:

(i) s_1 is a fixed line of both $\bar{\nu}$ and $\bar{\omega}$.

The proof is straightforward.

(ii) There exist points X, Y | s_1 such that $\omega = \nu XY$.

For i = 1, 2, 3, 4, $\bar{\beta}_i \bar{\alpha}_i^{-1}$ fixes the pencil of lines parallel to s_1 and has as its centre the pencil of perpendiculars to b_i and a_i. So $\bar{\beta}_i \bar{\alpha}_i^{-1}$ is a translation.

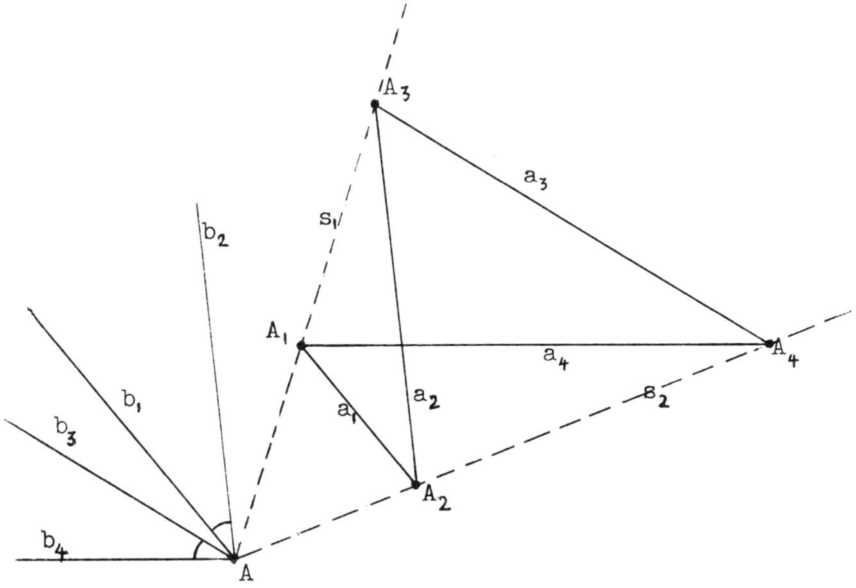

FIGURE 8

Since $\bar{\beta}_i \bar{\alpha}_i^{-1}$ is a translation for i = 1, 2, 3, 4, also $\bar{\nu}\bar{\omega}^{-1}$ must be a translation. So (ii) holds because of (i).

(iii) $\bar{\omega}$ is a quasirotation with centre A.

(iv) $\bar{\nu}$ is a quasirotation or a translation.

$\bar{\omega}$ has the fixed point A and $\bar{\nu}$ the fixed line s_1. Therefore the statements (iii) and (iv) are consequences of Lemma 2.2 (remember that $s_1 \mid s_1$ and that the elements of $\bar{S}$ in particular are quasirotations).

(v) $\bar{\nu}$ is either a shear with axis s_1 or a translation.

Assume that $\bar{\nu}$ is not a translation. Then it is a quasirotation fixing s_1 [(i) and (iv)] and there is a fixed point Q of $\bar{\nu}$ (the centre of the quasirotation). Since every fixed line of a quasirotation goes through its centre, we conclude that $Q \mid s_1$. Now there are two possibilities: either $\nu \in S(s_1)$ or $\bar{\nu}$ has exactly one fixed point, which is Q. Assume, if possible, that $\bar{\nu}$ has one and only one fixed point Q. In this case Q^{α_1} is the only fixed point of $\overline{\nu^{\alpha_1}}$, which also has the fixed line s_2. Therefore $Q^{\alpha_1} \mid s_2$, hence $Q^{\alpha_1} \mid s_1^{\alpha_1}$, s_2 and hence, because $s_1^{\alpha_1} \parallel s_2$, $s_1^{\alpha_1} = s_2$. The last equation forces $A_1 = A$, which is impossible for non-degenerate i-quadrilaterals (a_i, s_j). The conclusion is that $\nu \in S(s_1)$. So (v) holds.

(vi) $\omega \in S(s_1)$.

If $\bar{\nu}$ is a translation, then by (ii) so is $\bar{\omega}$ and, by (iii), it follows that $\omega = 1$. If $\nu \in S(s_1)$ then, by (ii), the collineation $\bar{\omega}$ has to be a glide-shear. But, by (iii), $\bar{\omega}$ is a quasirotation. The conclusion is that $\bar{\omega}$ must be a shear, which proves (vi) and the lemma 3.3.

THEOREM 3.4 (theorem of the antiorthological i-quadrilaterals). <u>Let</u> $(a_i; s_j)$ <u>and</u> (b_i, t_j) <u>be two non-degenerate</u>

i-*quadrilaterals with* $s_1 \mid t_2$ *and* $s_2 \mid t_1$ *and* $a_i \mid b_i$ *for* $i = 1, 2, 3$. *Then also* $a_4 \mid b_4$.

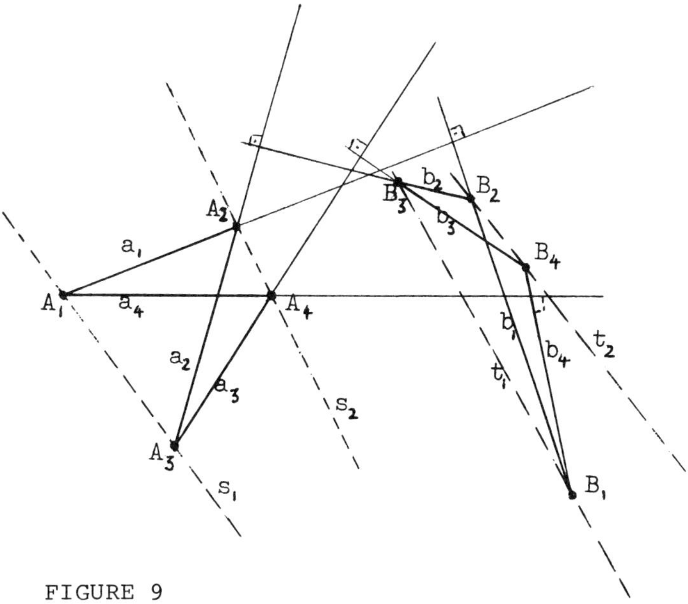

FIGURE 9

THEOREM 3.5 (theorem of the parallogical i-quadrilaterals). *Let* (a_i, s_j) *and* (b_i, t_j) *be non-degenerate* i-*quadrilaterals with* $s_j \parallel t_j$ *and* $a_i \parallel b_i$ *for* $j = 1, 2$ *and* $i = 1, 2, 3$. *Then also* $a_4 \parallel b_4$.

Theorems 3.4 and 3.5 are both consequences of Lemmas 3.2 and 3.3 of the i-quadrilaterals. This is obvious if the i-quadrilaterals in question each have a diagonal point. The case in which the diagonals are parallel can be reduced to the general case with diagonal point (see Götzky, 1970).

II.4. Coordinates[1]

In this section the results of section II.3 are used to prove that the group plane $E = (G, S_G)$ is a unitary-minkowskian plane. The method is to construct both the skew field which coordinatizes the plane and the hermitian form which represents the orthogonality relation of E.

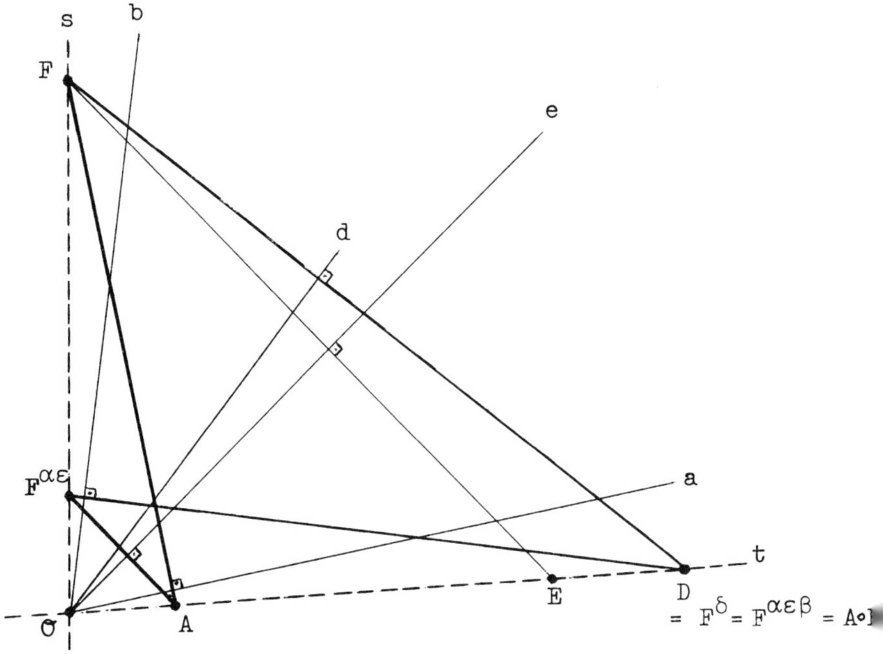

FIGURE 10

Let $\sigma \in S(s)$ with $s \mid s$ and $\varepsilon \in S(e)$ with $\varepsilon^2 = 1$. Further let $t = s^{\varepsilon}$ and $\tau = \sigma^{\varepsilon} \in S(t)$. O may be a point such that $O \mid e, s, t$. Let $F \mid s$ and $E = F^{\varepsilon}$, such that $F \neq O \neq E$.

1. Compare the introduction of coordinates in Wolff (1967).

From now on, for all the lines a, b, ... | O that satisfy $s \neq a \neq t$, $s \neq b \neq t$, ..., let α, β, ... represent elements of S(a), S(b), ... respectively, such that $s^{\alpha} = t$, $s^{\beta} = t$, ... (remember Axiom A.6). Further let $A := F^{\alpha}$, $B := F^{\beta}$, ...

Next let K be the set of all points incident with t, and define:

$A + B := AOB$ for all A, B $\in$ K,

$A \circ B := F^{\alpha\varepsilon\beta}$ for all A, B $\in$ K\{O},

$A \circ O := O =: O \circ A$ for all A $\in$ K.

Then K^{+} (standing for K with respect to the operation +) is an abelian group with unit O, in which A has the inverse $-A = OAO$. Further K° (representing K\{O} with respect to the operation $\circ$) is a not necessarily abelian group with unit $E = F^{\varepsilon}$, in which A has the inverse $A^{-} = F^{\varepsilon\alpha^{-1}\varepsilon}$ (remember that $\varepsilon^{2} = 1$ and hence $t^{\varepsilon} = s$).

Concerning the proof of these statements there is only one problem worth mentioning, and that is the associativity of K° which we now prove.

$A \circ B = A^{\varepsilon\beta}$; therefore $(A \circ B) \circ C = (A \circ B)^{\varepsilon\gamma} = A^{\varepsilon\beta\varepsilon\gamma}$. Let $D = B \circ C$, hence $D = F^{\delta} = F^{\beta\varepsilon\gamma}$. Obviously $\sigma' = \beta\varepsilon\gamma\delta^{-1}$ has the fixed points O and F. Since $O \neq F$ the reduction theorem yields $\sigma' \in S(s)$. Thus

$$A^{\varepsilon\beta\varepsilon\gamma} = A^{\varepsilon\sigma'\delta} = A^{\varepsilon\delta} = A \circ D;$$

hence

$$(A \circ B) \circ C = A \circ D = A \circ (B \circ C).$$

Since $(A + B) \circ C = (A + B)^{\varepsilon\gamma} = (AOB)^{\varepsilon\gamma} = A^{\varepsilon\gamma}OB^{\varepsilon\gamma}$ $= A\circ C + B\circ C$, K is a skew field if and only if

(i) $A \circ (B + C) = A\circ B + A\circ C$

also holds. Therefore K is a skew field if there is an antiautomorphism π of K.

CONSTRUCTION OF AN ANTIAUTOMORPHISM π OF K. Let a, b, ... be lines incident with O and let a*, b*, ... be the perpendicular lines from O onto a, b, ... respectively. The mapping

$$\pi: \begin{cases} K \longrightarrow K \text{ with } O \longrightarrow O \\ A \longrightarrow F^{\alpha^* O} = A^{*O} \quad \text{for} \quad A \in K^\circ \end{cases}$$

is involutory.[2] We shall see that π is an antiautomorphism of K with respect to the operations +, $\circ$.

First, let $D = A \circ B$. Considering figure 6 yields $\sigma' = \alpha\varepsilon^{-1}\beta\delta^{-1} \in S(s)$ and the second lemma of the i-quadrilaterals yields $\sigma'' = \alpha^*\delta^{*-1}\beta^*\varepsilon^{*-1} \in S(s)$ (Lemmas 3.2 and 3.3). Thus

$$\beta^*\varepsilon^{*-1}\alpha^*\delta^{*-1} \in S(s^{\alpha^*\delta^{*-1}}) = S(s).$$

---

2. Since $E(G, S_G)$ is a translation plane, the theorem of the parallogical i-quadrilaterals (Theorem 3.5) implies that the line reflection σ with axis s in the direction of t exists. Therefore $(A^*)^O = A^{\alpha^{-1}\alpha^* O} = A^{O\cdot O\alpha^{-1}\alpha^* O}$ $= (A^O)^{\sigma\alpha^{-1}\sigma\cdot\sigma\alpha^*\sigma}$ holds for all $A \in K^\circ$. If now $d \mid O$ such that $A^O = F^\delta$, then $\delta = \sigma\alpha\sigma$ and $\delta^* = \sigma\alpha^*\sigma$; hence $(A^*)^O =$ $(A^O)^{\delta^{-1}\delta^*} = (A^O)^*$. So $*O = O*$, and therefore $\pi^2 = *^2\cdot O^2 = 1$

But $\varepsilon^*O = \varepsilon$ and $\varepsilon^{*^{-1}} = \varepsilon^*$, hence $F^{\beta^*\varepsilon\alpha^*} = F^{\delta^*O}$, hence

$$D^\pi = F^{\delta^*O} = F^{\beta^*\varepsilon\alpha} = (F^{\beta^*O})^{\varepsilon\alpha^*O}.$$

If $\gamma \in S$ such that $F^{\alpha^*O} = F^\gamma$ and $c \mid O$ for $\gamma \in S(c)$, then $\alpha^*O\gamma^{-1} \in S(s)$. Hence

$$D^\pi = (F^{\beta^*O})^{\varepsilon\alpha^*O} = (F^{\beta^*O})^{\varepsilon\gamma} = B^\pi \circ A^\pi.$$

So π is at least an antiautomorphism of K°.

Second, let D = A + B. Consider then figure 11.

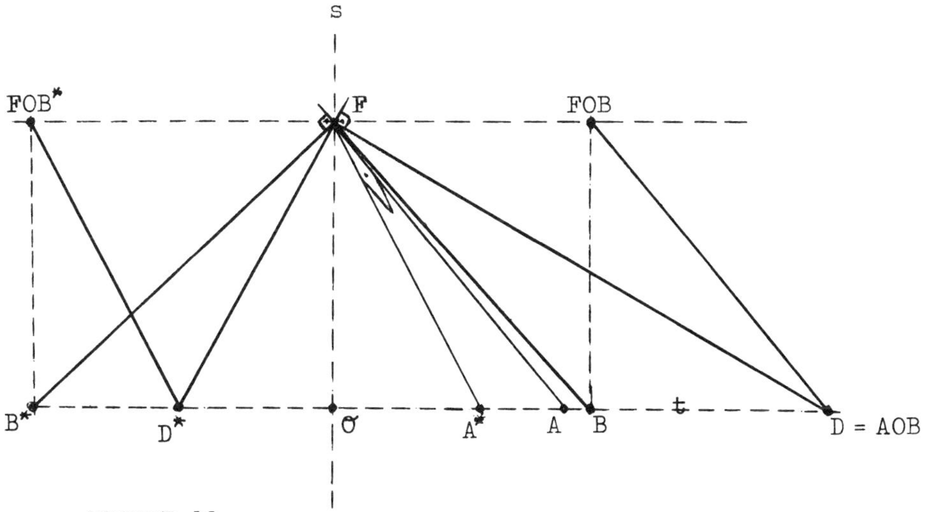

FIGURE 11

The theorem of the antiorthological i-quadrilaterals (Theorem 3.4) applied to the i-quadrilaterals with vertices F, B, FOB, D and F, B*, FOB*, D* yields $\overline{(FOB)D} \mid \overline{(FOB^*)D^*}$. Since $\overline{FA} \parallel \overline{(FOB)D}$ and $\overline{FA^*} \mid \overline{FA}$ we get

$$\overline{FA} \mid \overline{(FOB^*)D^*} \parallel \overline{FA^*}.$$

So F(FOB*) = A*D*, hence A*D* = OB*, hence D* = A*OB*. The last equation yields

$D^\pi = A^\pi O B^\pi$

which proves that π is an automorphism of K^+.

So π is an antiautomorphism of K, and the following theorem is proved:

THEOREM 4.1. $K = (K, +, \circ)$ _is a skew field with antiautomorphism_ π.

Next we introduce coordinates in $E = E(G, S_G)$ using K as skew field of coordinates.

Since E is a translation plane, there is for each point P_i of E exactly one pair $X_i, Y_i \in K$ such that

(ii) $P_i = X_i O Y_i^\varepsilon$.

Conversely every pair $X_i, Y_i \in K$ determines a point P_i by (ii). We have to show that every line considered as a set of points can be described by a linear equation.

First, we describe lines parallel to s or t.

Obviously

(1) lines parallel to s are described by equations $X = C$,

(2) lines parallel to t are described by equations $Y = C$.

Second, we describe lines incident with O. Let $a \mid O$ and $O \neq P_1, P_2 \mid a$. Consider figure 12. By the theorem of the parallogical i-quadrilaterals (Theorem 3.5) $\overline{X_1 Y_1^\varepsilon} \parallel \overline{X_2 Y_2^\varepsilon}$. Hence there is a line $\tilde{a}$ perpendicular to both the lines $\overline{X_1 Y_1^\varepsilon}$ and $\overline{X_2 Y_2^\varepsilon}$ and passing through O. Then $Y_i^\varepsilon = X_i^{\tilde{\alpha}^{-1}}$ for $i = 1, 2$, so $X_i = Y_i \circ \tilde{A}$.

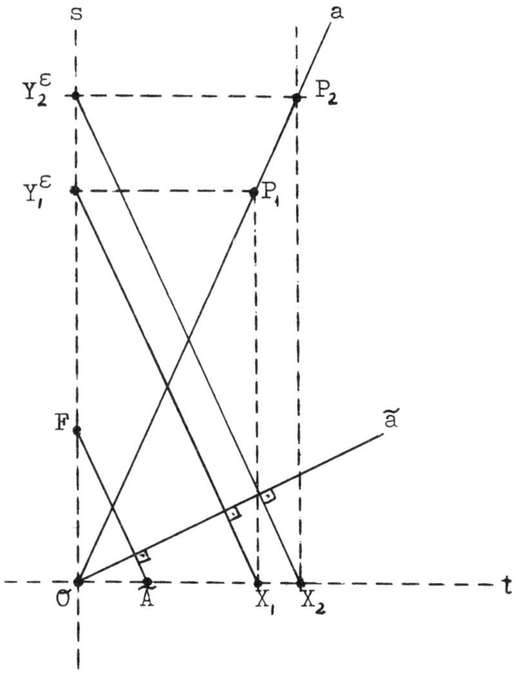

FIGURE 12

Thus

(3) lines a which are incident with O but different from s and t are described by equations $X = Y \circ \tilde{A}$.

Finally, let s, $t \neq a \mid O$ and $b \parallel a$. There is a point $Z \in K$ such that $b = a^{OZ^\varepsilon} = a^{Z^\varepsilon}$ (remember that E is a translation plane). If now $P = XOY^\varepsilon$ is a point of b, the above result yields:

$$X = Y^{Z^\varepsilon O} \circ \tilde{A} = (OZ^\varepsilon YZ^\varepsilon O) \circ \tilde{A} = (YZ^\varepsilon OZ^\varepsilon O) \circ \tilde{A}$$
$$= (YOO^{Z^\varepsilon O}) \circ \tilde{A} = (Y + O^{Z^\varepsilon O}) \circ \tilde{A} = Y \circ \tilde{A} + O^{Z^\varepsilon O} \circ \tilde{A}.$$

Let $\hat{B} := O^{Z^\varepsilon O} \circ \tilde{A}$; then $X = Y \circ \tilde{A} + \hat{B}$.

This proves

(4) lines b parallel neither to s nor to t are described by equations $X = Y \circ \tilde{A} + \hat{B}$.

Since K is a skew field a general line equation can be derived from (1) to (4):

(5) $X \circ A + Y \circ B + C = O$.

If now $V = V_3(K)$ is the 3-dimensional right vector space over the skew field K, it is easy to realize that the lines of our plane can be represented by the 1-dimensional subspaces of V which are different from $(O, O, E) \circ K$. Moreover, with $\mathcal{P} = PV$ the planes $\mathcal{P}_{(O,O,E)\circ K}$ and $E = E(G, S_G)$ are isomorphic. (Compare the introduction of coordinates in Artin, 1957.)

Next we describe the orthogonality of $E = E(G, S_G)$.

THEOREM 4.2. <u>Let</u> a, b <u>be lines of</u> $E = E(G, S_G)$ <u>represented by the vectors</u> (A_1, A_2, A_3) <u>and</u> (B_1, B_2, B_3). <u>Then</u> $a \mid b$ <u>if and only if</u>

$$(A_1^{\pi}, A_2^{\pi}, A_3^{\pi}) \circ \begin{pmatrix} O & E & O \\ E & O & O \\ O & O & O \end{pmatrix} \circ \begin{pmatrix} B_1 \\ B_2 \\ B_3 \end{pmatrix} = O.$$

PROOF FOR $A_2 = O$. Since a, b are lines of E, it is easy to see that the inequalities $A_1 \circ A_2 \neq O \neq B_1 \circ B_2$ hold. Now $A_2 = O$ means that $a \parallel s$ or, since $s \mid s$, that $a \mid s$. So $a \mid b$ if and only if $B_2 = O$. But assuming $A_2 = O$, the matrix equation holds if and only if $B_2 = O$.

PROOF FOR $A_2 \neq O$. Some considerations similar to those just used show that we may assume $B_2 \neq O$ without loss of generality. Let therefore $A_2 \neq O \neq B_2$.

The matrix equation holds if and only if $A_1^\pi \circ B_2 + A_2^\pi \circ B_1 = O$, hence if and only if $(A_1 \circ A_2^-)^\pi + B_1 \circ B_2^- = O$. Using the symbols out of (4), this last equation becomes transformed into $-(\tilde{A}^-)^\pi - \tilde{B}^- = O$, or $\tilde{A}^\pi = \tilde{B}^O$, or $\tilde{A}^* = \tilde{B}$. The last equation is equivalent to $\tilde{a} \mid \tilde{b}$ (see figure 12 and remember the meaning of * used for the construction π).

It remains to show that $\tilde{a} \mid \tilde{b}$ if and only if $a \mid b$.

Let $P_1 := X_1 O Y_1^\varepsilon \neq O \neq Q_2 := X_2 O Y_2^\varepsilon$ with $P_1 \mid a$ and $Q_2 \mid b$. To visualize this, refer to figure 12. We have $\tilde{a} \mid \tilde{b}$ if and only if $\overline{X_1 Y_1^\varepsilon} \mid \overline{X_2 Y_2^\varepsilon}$. Hence, applying the theorem of the antiorthological i-quadrilaterals to the quadrilaterals with vertices O, P_1, Y_1^ε, X_1 and O, Q_2, Y_2^ε, X_2 respectively, we find that $\tilde{a} \mid \tilde{b}$ if and only if $a \mid b$.

This completes the proof of 4.2.

MAIN THEOREM 4.3. *Let* f *be the* π-*hermitian form* (*right linear*, *since* $V_3(K)$ *is a right vector space*) *defined by the matrix*

$$\begin{pmatrix} O & E & O \\ E & O & O \\ O & O & O \end{pmatrix}.$$

Further, let $V = V_3(K, f, \pi)$.

Then $E(G, S_G) \simeq \mathcal{P}_{\text{Rad } V}$ *for* $\mathcal{P} = PV$ *and* $G = \bar{G} = U_3^*(K, f, \pi)$. *Hence, in particular*, $E(G, S_G)$ *is a unitary-minkowskian plane*.

PROOF. Taking Theorem 4.2 into consideration, we know already that $E(G, S_G) \simeq P_{\text{Rad } V}$. By Theorem 2.3, $G \simeq \bar{G}$ holds. We now compare the generated groups $(\bar{G}, \bar{S})$ and (PU*, PS(V)) (for the definition of the last group see Part I). Since PS(V) is the set of all axial collineations of $P_{\text{Rad } V}$ which preserve the orthogonality, we infer from Theorem 1.3 that identifying $E(G, S_G)$ and $P_{\text{Rad } V}$ the inclusion $\bar{S} \leq PS(V)$ becomes valid. Because now Axiom A.6 yields $PS(V) \leq \bar{S}$, we get $(G, S) \simeq (PU^*, PS(V))$. Since finally $U^* := U_3^*(K, f, \pi) \simeq PU^*$ the main theorem must hold.

It remains only to remark that if $U^* = U_3^*(K, f, \pi)$ is a restricted unitary group of the minkowskian type, then (PU*, PS(V)) satisfies the basic assumption and the axioms of section II.1.

REFERENCES

Artin, E. 1957. Geometric Algebra. Interscience Tracts, No. 3, Interscience Publishers, New York.

Bachmann, F. 1959. Aufbau der Geometrie aus dem Spiegelungsbegriff. Die Grundlehren der mathematischen Wissenschaften, Band 96. Erste Auflage 1959, zweite Auflage Berlin-Heidelberg-New York 1973.

Götzky, M. 1964. Eine Kennzeichnung der orthogonalen Gruppen unter den unitären Gruppen. Arch. d. Math. 15: 161-165.

—— 1965. Eine Kennzeichnung der unitären Gruppen über einem Schiefkörper der Charakteristik ≠ 2. Dissertation, Kiel.

—— 1970. Aufbau der unitär-minkowskischen Geometrie mit Hilfe von Quasispiegelungen. Habilitationsschrift, Kiel.

—— 1972. Mittelpunktsabbildungen in affinen Ebenen. Abh. Math. Sem. Hamburger Univ. 37: 133-146.

—— 1973. Der Satz von der dritten Quasispiegelung. Supplement 15 in Bachmann (1973).

Pickert, G. 1955. Projektive Ebenen. Springer, Berlin-Göttingen-Heidelberg.

Schütte, K. 1955. Ein Schliessungssatz für Inzidenz und Orthogonalität. Math. Ann. 129: 424-430.
Veblen, O. and J.W. Young. 1910, 1918. Projective Geometry, vols. 1, 2. Boston.
Wolff, H. 1967. Minkowskische und absolute Geometrie. Math. Ann. 171: 144-193.

THE LENZ-BARLOTTI TYPE III

Christoph Hering

1. INTRODUCTION

A finite projective plane admitting a collineation group G of Lenz type III is desarguesian. The first published proof of this theorem (see [5]) uses the classification of finite groups with a split BN-pair of rank 1. Therefore a complete presentation of this proof necessarily must be very long. Also, the proof is quite indirect. It involves the characterization of a large variety of groups, namely the groups PSL(2, q), Sz(q), PSU(3, q), groups of Ree type, and sharply doubly transitive groups, although in the end, because of geometric reasons, only the groups PSL(2, q) actually occur. Therefore it seems worth while to look for a more direct proof which always stays within the range of the 2-dimensional linear fractional groups. For planes of odd order such a proof actually exists, and it is the main purpose of this paper to describe this shorter, more economic way of argumentatic

Instead of the classification of finite groups with a split BN-pair of rank 1 we use here a system of three theorems on group spaces. At first glance, this system looks somewhat artificial. The main idea is to save for G a sufficient amount of geometric information without losing the possibility of applying induction. This is quite difficult and could only be achieved by partitioning the problem into three different cases. One of the three cases (see Theorem C) deals with groups in which all involutions are central. Theorem 2.1 and Corollary 2.2 list a few simple properties of such groups. Possibly one can find stronger results of this kind, which might lead to further simplifications. The proof presented here does not use any group-theoretical classification theorem apart from the classification of finite groups with dihedral Sylow 2-subgroups by Gorenstein and Walter [2] (which of course uses the theorem of Feit and Thompson). Even this theorem is not used in its full generality but only in a certain quite special situation (see Proposition 3.6).

Our notation is fairly standard. Let G be a group, a and b elements of G, and $\mathcal{A}$ as well as $\mathcal{B}$ subsets of G. Then $a^b = b^{-1}ab$, $[a, b] = a^{-1}b^{-1}ab$, and $[\mathcal{A}, \mathcal{B}] = \langle [a, b] \mid a \in \mathcal{A} \text{ and } b \in \mathcal{B} \rangle$. Also, $\mathcal{Z}G$ is the centre of G, $\mathcal{C}_G \mathcal{A}$ the centralizer of $\mathcal{A}$ in G, $\mathcal{N}_G \mathcal{A}$ the normalizer of $\mathcal{A}$ in G, G' the commutator subgroup of G, $G^{\#}$ the set consisting of all non-trivial elements in G, O(G) the largest normal subgroup of odd order of G, and $\Omega_2(G)$ the subgroup generated by all involutions in G. An involution is always an element of order 2. A dihedral group is a group with two generators a and b and the relations

$$a^n = b^2 = 1, \quad bab = a^{-1},$$

where $n \geq 2$.

Suppose that in addition to the group G we have a set Ω and a binary operation $\Omega \times G \to \Omega$, such that $\alpha^{gh} = (\alpha^g)^h$ and $\alpha^1 = \alpha$ for all $\alpha \in \Omega$ and $g, h \in G$. Then we call the pair (Ω, G) together with the given binary operation a group space. Furthermore, we denote $\Omega(\mathfrak{A}) = \{\omega \in \Omega \mid \omega^a = \omega \text{ for all } a \in \mathfrak{A}\}$, and $G_{\alpha_1 \ldots \alpha_n} = \{g \in G \mid \alpha_i^g = \alpha_i \text{ for } 1 \leq i \leq n\}$, if $\alpha_1, \ldots, \alpha_n \in \Omega$.

2. GEOMETRIC AND GROUP-THEORETICAL TOOLS

In this section we present the most specific group-theoretical and geometric results which will be used in our proof of the Type III Theorem.

THEOREM 2.1. Let G be a finite group containing a subgroup Z such that

$$\Omega_2(G) \leq Z \leq \mathfrak{Z}G,$$

and let U be a subgroup of G containing Z. Choose a subgroup N of Z maximal with respect to the property that

$$\Omega_2(U/N) \leq Z/N.$$

Then the following statements hold:

(a) Z/N *is an elementary abelian* 2-*group*.

(b) *If* $1 \neq x \in Z/N$, *then there exists an element* $y \in U/N - Z/N$ *such that* $y^2 = x$.

(c) *The number* r *of conjugacy classes of involutions in* G/Z *intersecting* U/Z *non-trivially is at least* $|Z/N| - 1$. *In particular*, $r \geq 3$, *unless the Sylow* 2-*subgroups of* U/Z *are cyclic or dihedral*.

(d) $|Z/N| \leq |U/Z|$.

PROOF. Denote $\bar{G} = G/N$, $\bar{U} = U/N$, and $\bar{Z} = Z/N$. Let A be a subgroup of odd order of $\bar{Z}$. As $\bar{Z} \leq \mathfrak{Z}\bar{U}$, $A \trianglelefteq \bar{U}$. If xA is an involution in $\bar{U}/A$, then $|\langle x, A\rangle| = 2|A|$ so that xA contains an involution $\tilde{x}$ and $xA = \tilde{x}A \in \bar{Z}/A$. So $\Omega_2(\bar{U}/A) \leq \bar{Z}/A$, and by maximality of N we obtain $A = 1$. Suppose now that $\bar{Z}$ contains an element z such that $y^2 \neq z$ for all $y \in \bar{U} - \bar{Z}$. For each involution $x\langle z\rangle$ in $\bar{U}/\langle z\rangle$ the group $\langle x, z\rangle$ is an abelian 2-group containing the cyclic group $\langle z\rangle$ as a subgroup of index 2. Hence either $\langle x, z\rangle$ is cyclic or $\langle x, z\rangle = \langle \bar{x}\rangle \times \langle z\rangle$, where $\bar{x}$ is an involution. In the first case there exists an element y such that $y^2 = z$ and $\langle x, z\rangle = \langle y\rangle$. As $y^2 = z$, we have $y \in \bar{Z}$, so that $x\langle z\rangle \in \bar{Z}/\langle z\rangle$. In the second case $\bar{x} \in \bar{Z}$ by assumption. But this implies that $\langle x, z\rangle = \langle \bar{x}, z\rangle \leq \bar{Z}$ and $x\langle z\rangle \in \bar{Z}/\langle z\rangle$. So $\Omega_2(\bar{U}/\langle z\rangle) \leq \bar{Z}/\langle z\rangle$, and by maximality of N we obtain $\langle z\rangle = 1$, which proves (b).

Suppose now that $\bar{Z}$ contains an element a of order 4. By (b), there exists an element $y \in \bar{U} - \bar{Z}$ such that $y^2 = a^2$. As $a \in \mathfrak{Z}\bar{U}$, the group $\langle y, a\rangle$ is an abelian group of order 8. It contains two cyclic groups of order 4, namely $\langle y\rangle$ and $\langle a\rangle$. Therefore $\langle y, a\rangle$ is the direct product of a cyclic group of order 4 and a cyclic group of order 2. But $\langle y, a\rangle \cap \bar{Z} = \langle a\rangle$. Hence we have involutions in $\langle y, a\rangle$, which do not lie in $\bar{Z}$, a contradiction. So $\bar{Z}$ is an elementary abelian 2-group.

Let z_1 and z_2 be two different non-trivial elements in $\bar{Z}$. By (b), there exist elements u_1 and u_2 in $\bar{U} - \bar{Z}$ such that $u_1^2 = z_1$ and $u_2^2 = z_2$. Clearly $u_1\bar{Z}$ and $u_2\bar{Z}$ are involutions in $\bar{U}/\bar{Z}$. Suppose that these involutions are conjugate in $\bar{G}/\bar{Z}$. Then there exists an element $g \in \bar{G}$ such that $g^{-1}u_1g\bar{Z} = u_2\bar{Z}$. But this implies that $g^{-1}u_1g = u_2\bar{z}$ for a suitable element $\bar{z} \in \bar{Z}$, and hence $z_2 = u_2^2 = u_2^2\bar{z}^2 = (u_2\bar{z})^2 = (g^{-1}u_1g)^2 = g^{-1}u_1^2g = g^{-1}z_1g = z_1$, a contradiction. So the number of conjugacy classes of involutions in $\bar{G}/\bar{Z}$ intersecting $\bar{U}/\bar{Z}$ non-trivially is at least as large as the number of non-trivial elements in $\bar{Z}$. This number is at least three, unless $|\bar{Z}| = 2$ and $\bar{U}$ contains exactly one involution, in which case the Sylow 2-subgroups of U/Z are cyclic or dihedral. Thus the proof of (c) is complete, and as an immediate consequence we obtain (d). Finally, we consider the special case U = G. This leads to the following corollary:

COROLLARY 2.2. *Let* G *be a finite group containing a subgroup* Z *such that* $\Omega_2(G) \leq Z \leq \mathfrak{Z}G$. *If* G/Z *has at most two classes of involutions, then the Sylow* 2-*subgroups of* G/Z *are cyclic or dihedral*.

LEMMA 2.3 (Zassenhaus [11]). *Let* τ *be an involution acting on a finite group* G. *If* $\mathcal{L}_G \tau = 1$, *then* G *is abelian and* $g^\tau = g^{-1}$ *for all* $g \in G$.

PROOF. If $x \in G$, then $(x^{-1}x^\tau)^\tau = (x^\tau)^{-1}x = (x^{-1}x^\tau)^{-1}$, i.e. τ inverts the element $x^{-1}x^\tau$. On the other hand, our assumption implies that the map $x \mapsto x^{-1}x^\tau$ for all $x \in G$ is one-to-one and therefore a map of G onto itself. Thus τ inverts each element of G. This implies that $xy = ((xy)^{-1})^{-1} = ((xy)^\tau)^{-1} = (x^\tau y^\tau)^{-1} = (x^{-1}y^{-1})^{-1} = yx$, for all $x, y \in G$, so that G is abelian.

BRAUER-WIELANDT THEOREM 2.4 (see Wielandt [10]). *Let* $E = \{1, e_1, e_2, e_3\}$ *be an elementary abelian group of order* 4 *acting on a group* G *of finite odd order*. *Then*

$$|\mathcal{L}_G e_1| \; |\mathcal{L}_G e_2| \; |\mathcal{L}_G e_3| = |G| \; |\mathcal{L}_G E|^2.$$

THEOREM 2.5 (Gorenstein and Walter [2]). *A finite group with dihedral Sylow* 2-*subgroups satisfies one of the following conditions*:

(i) G/O(G) *is isomorphic to a subgroup of* PΓL(2, q) *containing* PSL(2, q) *for a suitable odd prime power* q.

(ii) G/O(G) *is isomorphic to the alternating group* A_7.

(iii) $G/O(G)$ *is isomorphic to a Sylow 2-subgroup of* G.

LEMMA 2.6 (see [3, Hilfssatz 6]). *Let* G *be a finite doubly transitive permutation group which contains involutions fixing 2 letters but no involutions fixing more than 2 letters. Then the Sylow 2-subgroups of* G *are dihedral groups or quasidihedral groups*.

LEMMA 2.7. *In* PSL(2, q), q *odd, the normalizer of each elementary abelian subgroup* E *of order 4 is transitive on* $E - \{1\}$.

PROOF. Let q be an odd prime power and $G \simeq PSL(2, q)$. Then G is not 2-nilpotent. Hence by the Transfer Theorem of Frobenius [7, p. 436, Satz 5.8] G contains a 2-group E such that $\mathfrak{N}_G E/\mathfrak{C}_G E$ is not a 2-group. As E is a dihedral group, it follows that E is elementary abelian of order 4. Hence the order of $\mathfrak{N}_G E/\mathfrak{C}_G E$ is divisible by 3, so that $\mathfrak{N}_G E$ is transitive on the set of non-zero elements of E. We finish the proof by showing that the automorphism group of G acts transitively on the set consisting of all subgroups of G which are isomorphic to E. Let $F \leq G$ be a second elementary abelian subgroup of order 4. As all involutions are conjugate in G, we can assume without loss of generality that $E \cap F$ contains an involution α. Let C be the centralizer of α in PGL(2, q). Then C is a dihedral group and $E/\langle\alpha\rangle$ and $F/\langle\alpha\rangle$ are involutions in $(C \cap G)/\langle\alpha\rangle$. These involutions must be conjugate in $C/\langle\alpha\rangle$, as E and F are not cyclic.

LEMMA 2.8 (Baer [1]). *An involutory automorphism of a projective plane of finite odd order is either a homology or a Baer involution.*

LEMMA 2.9 (Ostrom [8] and Lüneburg). *Let* G *be a group of automorphisms of a projective plane* $(\mathcal{P}, \mathcal{L})$, $a \in \mathcal{L}$, $b \in \mathcal{L} - a$, $A \in b - a$, $B \in a - b$. *If* α *and* β *are involutions, then* $\alpha\beta$ *is the unique involution in* $G(a \cap b, AB)$.

PROOF. As the commutator $[\alpha, \beta]$ fixes each point on a and b, it is trivial. Hence $\alpha\beta$ is an involution. This involution fixes a and exactly two points on a. A Baer involution fixes at least three points on each of its fixed lines. Therefore the product $\alpha\beta$ is a perspectivity, which clearly must have $a \cap b$ as centre and AB as axis. Finally, if γ is any involution in $G(a \cap b, AB)$, then by the part of the theorem which we have already proved, $\alpha(\alpha\beta)$ and $\alpha\gamma$ lie in $G(B, b)$. On the other hand $\alpha(\alpha\beta)(\alpha\gamma)^{-1}$ fixes each point on AB and of course also each point on b, which implies that $\alpha(\alpha\beta) = \alpha\gamma$. Hence $\alpha\beta$ is unique.

3. THREE THEOREMS ON GROUP SPACES

Let (Ω, G) be a finite group space with the property:

(*) <u>For each</u> $\alpha \in \Omega$ <u>the stabilizer</u> G_α <u>contains a normal subgroup</u> N_α <u>which is sharply transitive on</u> $\Omega - \{\alpha\}$, <u>such that for</u> $\alpha \in \Omega$ <u>and</u> $g \in G$ <u>we have</u> $(N_\alpha)^g = N_{\alpha^g}$.

In this section we derive some properties of (Ω, G) which will be useful for the proof of our main theorem. We introduce the following notation:

$S = \langle N_\alpha \mid \alpha \in \Omega\rangle$.

ϕ is the representation of G on Ω determined by the group space (Ω, G). K is the kernel of ϕ. If U is a subgroup or an element of G, then we denote $U^\phi = \bar{U}$.

If U is a subgroup or an element of G, then $\Omega(U)$ is the set of fixed points of U.

If U is a subgroup or an element of G and $|\Omega(U)| \geq 3$, then $S(U) = \langle N_\alpha \cap \mathcal{L}_G U \mid \alpha \in \Omega(U)\rangle$ and $K(U) = \{x \in S(U) \mid \xi^x = \xi \text{ for all } \xi \in \Omega(U)\}$.

An involution in G is called an n-<u>involution</u> if it fixes exactly n points in Ω.

PROPOSITION 3.1. <u>If</u> $U \leq G_{\alpha\beta}$, <u>where</u> α <u>and</u> β <u>are two different points in</u> Ω, <u>then</u> $N_\alpha \cap \mathcal{L}_G U = \{x \in N_\alpha \mid \beta^x \in \Omega(U)\}$. <u>In particular</u>, $|N_\alpha \cap \mathcal{L}_G U| = |\Omega(U)| - 1$.

PROOF. If $x \in N_\alpha \cap \mathcal{L}_G U$, then $\beta^{xu} = \beta^{ux} = \beta^x$ for all $u \in U$, so that $\beta^x \in \Omega(U)$. Assume now that $x \in N_\alpha$ and $\beta^x \in \Omega(U)$. Then for $u \in U$ the commutator $xux^{-1}u^{-1}$ fixes β. On the other hand it lies in N_α, as U normalizes N_α. Hence $xux^{-1}u^{-1} = 1$ and $x \in N_\alpha \cap \mathcal{L}_G U$.

Let $U \leq G$ such that $|\Omega(U)| \geq 3$. Then obviously $\mathcal{N}_G U$ leaves invariant $\Omega(U)$ so that we obtain a new group space $(\Omega(U), \mathcal{N}_G U)$. The following statement is very important because it allows us to apply induction in many instances. It says that $(\Omega(U), \mathcal{N}_G U)$ again is a group space with property (*), and that in this group space $S(U)$ takes the place of S.

INDUCTION LEMMA 3.2. Let U be a subgroup of G fixing at least three points and denote $\hat{N}_\alpha = N_\alpha \cap \mathcal{L}_G U$ for all $\alpha \in \Omega(U)$. Then

(a) $\hat{N}_\alpha \trianglelefteq (\mathcal{N}_G U)_\alpha$, $\hat{N}_\alpha$ is sharply transitive on $\Omega(U) - \{\alpha\}$ for all $\alpha \in \Omega(U)$ and

(b) $(\hat{N}_\alpha)^g = \hat{N}_{\alpha^g}$ for all $\alpha \in \Omega(U)$ and $g \in \mathcal{N}_G U$.

PROOF. Statement (a) can easily be derived from Proposition 3.1. Also, $g \in \mathcal{N}_G U$ implies $(\hat{N}_\alpha)^g = (N_\alpha \cap \mathcal{L}_G U)^g$ $= (N_\alpha)^g \cap (\mathcal{L}_G U)^g = N_{\alpha^g} \cap \mathcal{L}_G U = \hat{N}_{\alpha^g}$.

Note that the property (*) also is inherited by each group space $(\Omega, G/N)$ where $N \trianglelefteq G$ and $N \leq K$. A further consequence of Proposition 3.1 is

PROPOSITION 3.3. $[K, S] = 1$.

PROPOSITION 3.4. Assume that $\bar{G}$ contains a 2-involution.

(a) If α and β are two different points in Ω, then $\bar{G}_{\alpha\beta}$ contains exactly one 2-involution.

(b) <u>All</u> 2-<u>involutions in</u> $\bar{G}$ <u>are conjugate</u>.

(c) <u>If the group</u> $\bar{G}$ <u>contains an involution fixing more than two points</u>, <u>then it contains an involution whose number of fixed points is at least</u> $(|\Omega| - 1)^{\frac{1}{2}} + 1$.

PROOF. Let t be an involution in $\bar{G}$ fixing two points α and β in Ω, but no further point. Then t normalizes $\bar{N}_\alpha$, and $\mathcal{L}_{\bar{N}_\alpha} t = 1$ by Proposition 3.1. The Zassenhaus Lemma 2.3 now implies that t inverts each element of $\bar{N}_\alpha$. Thus, for each further 2-involution t* in $\bar{G}_{\alpha\beta}$ we have $\bar{N}_\alpha \subseteq \mathcal{L}_{\bar{G}} tt^*$, so that tt* = 1 and t = t*, again by Proposition 3.1. As clearly (Ω, G) is doubly transitive, we obtain (a) and (b). Suppose now that $\bar{G}_{\alpha\beta}$ contains an involution u fixing more than two points. By (a) the 2-involution t is central in $\bar{G}_{\alpha\beta}$, so that $\langle t, u\rangle$ is an elementary abelian group of order 4. This group acts on $\bar{N}_\alpha$, where $|\bar{N}_\alpha| \equiv |\Omega| - 1 \equiv |\Omega(t)| - 1 \equiv 1 \pmod 2$. Hence $|\mathcal{L}_{\bar{N}_\alpha} u| \cdot |\mathcal{L}_{\bar{N}_\alpha} tu| = |\bar{N}_\alpha| = |\Omega| - 1$, by the Brauer-Wielandt Theorem 2.4, so that $|\mathcal{L}_{\bar{N}_\alpha} u| \geq (|\Omega| - 1)^{\frac{1}{2}}$ or $|\mathcal{L}_{\bar{N}_\alpha} tu| \geq (|\Omega| - 1)^{\frac{1}{2}}$. Proposition 3.1 now implies (c).

PROPOSITION 3.5. <u>If</u> S <u>contains an element fixing exactly two points</u>, <u>then</u> S = S'. <u>If</u> $|\Omega|$ <u>is even</u>, <u>then</u> S <u>does not contain a normal subgroup of index</u> 2, <u>and each element in</u> S <u>induces an even permutation of</u> Ω.

PROOF. Assume that S contains an element a such that $|\Omega(a)| = 2$. Let $\alpha \in \Omega(a)$. Then $\mathcal{L}_{N_\alpha} a = 1$ by 3.1. Hence the map $x \mapsto [x, a]$ for all $x \in N_\alpha$ is a one-to-one map of N_α onto itself, so that $N_\alpha = [N_\alpha, a] \subseteq S'$. As G is doubly transitive, it follows that $S = S'$. Assume now that $|\Omega|$ is even. Then the groups N_α for $\alpha \in \Omega$ have odd order, which immediately implies the above statement.

PROPOSITION 3.6. _If the Sylow 2-subgroups of $\bar{S}$ are cyclic or dihedral groups and $4 \leq |\Omega| \equiv 0 \pmod 2$, then there exists an odd prime power q such that $|\Omega| = q + 1$ and $S \cong SL(2, q)$ or $PSL(2, q)$._

PROOF. As $\bar{S}$ is doubly transitive, each non-trivial normal subgroup of $\bar{S}$ is transitive on Ω and hence has even order. Therefore $O(\bar{S}) = 1$. Furthermore, $\bar{S}$ is not a 2-group, as $|\Omega|(|\Omega| - 1) \mid |\bar{S}|$. Hence the Sylow 2-subgroups of $\bar{S}$ are not cyclic and by a classification theorem of Gorenstein and Walter (see 2.5), $\bar{S}$ is isomorphic to a subgroup of $P\Gamma L(2, q)$ containing $PSL(2, q)$ for a suitable odd prime power q, or to the alternating group A_7. We prove that the last case does not arise. Suppose that $\bar{S} \cong A_7$ and denote $N = \bar{N}_\alpha$ and $n = |N|$ for some $\alpha \in \Omega$. Then $(n + 1)n \mid 7!/2$, and in particular $n < 50$, so that $|\bar{S}_\alpha| > 49$. Also, N is a subgroup of odd order of A_7. We consider the original permutation representation of A_7 on 7 letters. Assume at first that N has more than one non-trivial orbit. As each non-trivial orbit has at least length 3, we then have two orbits of length 3 and one of length 1, so that $|N| \mid 9$, $|\Omega| \le 10$, and $|\bar{S}_\alpha| \ge 252$. On the other hand, $\bar{S}_\alpha$ normalizes N and hence permutes the orbits of N among themselves. Hence $\bar{S}_\alpha$ contains a normal subgroup of index at most 2 fixing all three orbits of N. This subgroup is isomorphic to a subgroup of $S_3 \times S_3$ so that $|\bar{S}_\alpha| \mid 6^2 \cdot 2 < 252$, a contradiction. Assume now that N has exactly one non-trivial orbit $\mathcal{O}$. Then $|\mathcal{O}| = 3, 5$, or 7. The first case is impossible, as $|\bar{S}| > 6!$ so that $|\Omega| \ge 8$ and $n \ge 7$. If $|\mathcal{O}| = 5$, then $n \mid |S_5|$, so that $n \mid 15$ and actually $n = 15$, again because of the inequality $n \ge 7$. But $n = 15$ implies $16 = n + 1 \mid 7!/2$, a contradiction. So $|\mathcal{O}| = 7$. As the number of Sylow 7-subgroups in N is $\equiv 1 \pmod 7$ and smaller than 50/7, N must contain a normal Sylow 7-subgroup Y. Now Y is characteristic in N and hence normal in $\bar{S}_\alpha$. But this implies that $49 < |\bar{S}_\alpha| \mid 21$. Therefore there only remains the possibility that $\bar{S}$ is isomorphic to a subgroup of $P\Gamma L(2, q)$ containing $PSL(2, q)$, where $q = p^m$ for some odd prime p. Again, let $\alpha \in \Omega$ and $N = \bar{N}_\alpha$. Furthermore, let $\tilde{S}$ be the image of $PSL(2, q)$ in $\bar{S}$, $\tilde{N} = N \cap \tilde{S}$ and $\tilde{N}_p$ a Sylow p-subgroup of $\tilde{N}$. Note that $\bar{S}/\tilde{S}$ is abelian. On the other hand $\bar{S}$ does not contain any normal subgroup of index 2 because of Proposition 3.5. Hence $|\bar{S}/\tilde{S}| \equiv 1 \pmod 2$ and $|\bar{S}/\tilde{S}| \mid m$. Suppose at first that $\tilde{N} = 1$. Then N is abelian and $\mathcal{L}_{\tilde{S}_\alpha} N = N$ as N is sharply transitive on $\Omega - \{\alpha\}$. Also, N and $\tilde{S}_\alpha$ are two normal subgroups of $\bar{S}_\alpha$ with trivial intersection, so that

$[N, \tilde{S}_\alpha] = 1$ and $\tilde{S}_\alpha \le \mathcal{C}_{\tilde{S}_\alpha} N \cap \tilde{S} = N \cap \tilde{S} = 1$. But this leads to

$$m + 1 \ge |N| + 1 = |\Omega| \ge |\tilde{S}:\tilde{S}_\alpha| = |\tilde{S}|$$
$$= (q + 1)q(q - 1)/2.$$

This contradiction implies that $\tilde{N} \ne 1$. Suppose now that $\tilde{N}_p = 1$. Then $|\tilde{N}| \le (q + 1)/2$ and $|\mathcal{N}_{\tilde{S}}\tilde{N}| \le q + 1$ (see, for example, Huppert [7, p. 213, Hauptsatz 8.27]). As $\tilde{S}_\alpha \le \mathcal{N}_{\tilde{S}}\tilde{N}$, we obtain the inequality

$$m(q + 1)/2 + 1 \ge |N| + 1 = |\Omega| \ge |\tilde{S}:\tilde{S}_\alpha| \ge q(q - 1)/2,$$

so that $(m + 1)(q + 1) \ge q^2 - 1$ and $m + 1 \ge q - 1 = p^m - 1 \ge 3^m - 1$. Hence $m = 1$, $q \le 3$, and $|N| \le 2$, a contradiction. Therefore $\tilde{N}_p \ne 1$. By Sylow's theorem, $\tilde{N}_p$ is contained in a Sylow p-subgroup U of $\tilde{S}$. U is abelian and hence contained in $\mathcal{N}_{\tilde{S}}\tilde{N}_p$. As $\tilde{N}_p$ fixes exactly one point in Ω, namely α, we have $U \le \mathcal{N}_{\tilde{S}}\tilde{N}_p \le \tilde{S}_\alpha$. Since $\tilde{N}_p$ does not fix any point except α, the same holds for U. Hence even $\mathcal{N}_{\tilde{S}}U \le \tilde{S}_\alpha$. Now $\mathcal{N}_{\tilde{S}}U$ is maximal in $\tilde{S}$. Also $\tilde{S}$, being a non-trivial normal subgroup of a doubly transitive group, is transitive on Ω. Hence clearly $\tilde{S}_\alpha < \tilde{S}$, so that actually $\tilde{S}_\alpha = \mathcal{N}_{\tilde{S}}U$. This implies that $|\Omega| = q + 1$ and that $|\tilde{S}_{\alpha\beta}| = (q - 1)/2$ for $\alpha, \beta \in \Omega$ and $\alpha \ne \beta$. If $q > 3$, then Proposition 3.5 implies that $S' = S$. Hence in this case $\bar{S} = \tilde{S}$ and S is isomorphic to a homomorphic image of the representation group of $\bar{S}$. (Note that $S \cap K \le \mathcal{Z}S$ because of Proposition 3.3.) By Schur [9, pp. 119-120] it follows that $S \cong SL(2, q)$ or $S \cong PSL(2, q)$ except, possibly, if

$q = 9$, in which case the Schur multiplier of $\bar{S}$ has order 6. However, we also know that $S \cap K$ is a p-group: Let Y be the Sylow p-subgroup of $S \cap K$, and P a Sylow p-subgroup of S_α containing Y. Then $P = Y \times N_\alpha$. As $|S:S_\alpha| = q + 1$, P actually is a Sylow p-subgroup of S. Hence by a theorem of Gaschütz (see [7, p.121, Hauptsatz 17.4]) there exists a subgroup $\hat{S}$ of S such that $S = \hat{S}Y$ and $\hat{S} \cap Y = 1$. As Y is central, $\hat{S} \trianglelefteq S$, and $S/\hat{S} \cong Y$. But $S' = S$ now implies $Y = 1$. So $S \cong SL(2, q)$ or $PSL(2, q)$, whenever $S' = S$.

There still remains the case $q = 3$. In this case $\bar{S} \cong A_4$. Let T be a Sylow 2-subgroup of S. Then $\bar{T}$ is the subgroup of order 4 in $\bar{S}$, so that $\bar{T} \trianglelefteq \bar{S}$ and $T(S \cap K) \trianglelefteq S$. Now the normalizer of T contains T and the central subgroup $(S \cap K)$, hence the product $T(S \cap K)$, so that T is normal in $T(S \cap K)$. This implies that T is characteristic in $T(S \cap K)$ and therefore normal even in S. Hence TN_α is a group and as T is transitive, we have $S = TN_\alpha$. Because of Sylow's theorem, $3 \nmid |S \cap K|$, so that $S \cap K \leq T$. Clearly $T' \leq S \cap K$. Also, by a lemma of Zassenhaus (see [7, p. 350, Satz 13.4]) $T/T' = (S \cap K)/T' \times T_1/T'$, where T_1/T' is an elementary abelian group of order 4, such that T_1 is normalized by N_α and N_α acts non-trivially on T_1/T'. Here T_1 is transitive on Ω, so that $S = N_\alpha T_1$, $T_1 = T$, and $T' = S \cap K$. Thus T is generated by two elements a and b. We now determine the order of the commutator subgroup T'.

Let $x, y \in T$ and consider the commutator $[x, y]$. If $x \in S \cap K$, then $[x, y] = 1$. So we can assume that x lies in $T - (S \cap K)$. As N_α is transitive on $T/S \cap K - \{1\}$, there exists an element $u \in N_\alpha$ such that $x^u = a \cdot z$, where $z \in S \cap K$. Furthermore, y^u has the form $a^r b^s z'$, where $r, s \in \{1, 2\}$ and $z' \in S \cap K$. Hence

$$[x, y] = [x, y]^u = [x^u, y^u] = [az, a^r b^s z'] = [a, b^s].$$

Therefore $T' = \langle [a,b] \rangle$ and $|T'| \leq 2$, as $[a,b]^2 = [a,b^2] = 1$.

THEOREM A. *Assume that there exists a subgroup* U *in* SK *containing* K *as a normal subgroup of index* 2 *such that* $|\Omega(U)| = q + 1$ *and* $S(U)/K(U) \cong PSL(2, q)$ *for some odd prime power* q. *Then* $|\Omega| > q^2 + 1$.

PROOF. See [6, pp. 448-452].

THEOREM B. *Assume that* (Ω, G) *has the following properties*:

(a) *If* $\alpha, \beta \in \Omega$ *and* $\alpha \neq \beta$, *then* $G_{\alpha\beta}$ *contains exactly three involutions. One of them fixes all points in* Ω *while the remaining two don't fix any point different from* α *and* β. *All three are contained in* $\mathfrak{Z}G_{\alpha\beta}$ *but at least one is not contained in* $\mathfrak{Z}G_{\{\alpha,\beta\}}$.

(b) $2 < |\Omega| \equiv 2 \pmod 4$.

Then $S \cong SL(2, q)$ *and* $|\Omega| = q + 1$ *for a suitable prime power* q.

PROOF. Let (Ω, G) be a counterexample of smallest possible degree. Because of Proposition 3.5, each element of S induces an even permutation of Ω. By Proposition 3.6 S contains more than oneinvolution so that S - K contains an involution t. As $\bar{t}$ is an even permutation and $|\Omega| \equiv 2 \pmod 4$, t fixes at least two points α and β, and $\Omega(t) = \Omega(\bar{t}) = \{\alpha, \beta\}$ by (a). Suppose that $\bar{S}$ contains an involution fixing more than 2 points. Then because of Proposition 3.4 (c),

the stabilizer $\bar{S}$ contains an involution u such that $|\Omega(u)| \geq (|\Omega| - 1)^{\frac{1}{2}} + 1$. Let U be the pre-image of $\langle u \rangle$ in G. We consider the group space $(\Omega(U), \mathcal{L}_G U)$. As the three involutions in $G_{\alpha\beta}$ are central, they are contained in $\mathcal{L}_G U$. Let E be the group generated by these involutions and suppose that $E \leq \mathcal{Z}(\mathcal{L}_G U)_{\{\alpha,\beta\}}$. In the doubly transitive group space $(\Omega(U), \mathcal{L}_G U)$, there exists an element x interchanging α and β. As x centralizes E it follows that $G_{\{\alpha,\beta\}} = \langle x, G_{\alpha\beta} \rangle$ centralizes E, which is impossible by (a). So $E \leq \mathcal{Z}(\mathcal{L}_G U)_{\alpha\beta}$ but $E \not\leq \mathcal{Z}(\mathcal{L}_G U)_{\{\alpha,\beta\}}$. Also $u \in \bar{S}$, so that u is an even permutation of Ω and hence $|\Omega(U)| \equiv |\Omega| \equiv 2 \pmod 4$. Therefore we can apply induction to $(\Omega(U), \mathcal{L}_G U)$. It follows that $S(U) \cong SL(2, \bar{q})$ and $|\Omega(U)| = \bar{q} + 1$ for a suitable prime power $\bar{q}$. Because $\bar{q} + 1 = |\Omega(U)| \geq (|\Omega| - 1)^{\frac{1}{2}} + 1$, we obtain a contradiction to Theorem A.

So all involutions in $\bar{S}$ have at most 2 fixed points. As they induce even permutations on Ω and $|\Omega| \equiv 2 \pmod 4$, they actually all have exactly 2 fixed points. Hence by

Lemma 2.6 the Sylow 2-subgroups of $\bar{S}$ are dihedral groups or quasidihedral groups. Also, we can obtain information about the involutions in S. Clearly the kernel K contains exactly one involution by (a). If S contains only one involution, then we are finished by Proposition 3.6. So we may assume that S contains an involution t which is not contained in K. Then by the above t has two fixed points, w.l.o.g. α and β. Let E be the elementary abelian subgroup of order 4 generated by all involutions in $G_{\alpha\beta}$. As E is not central in $G_{\{\alpha,\beta\}}$, there exists an element $x \in G_{\{\alpha,\beta\}}$ such that $E = \langle t, t^x \rangle$. Hence $E \leq S$ and therefore $S \cap K \geq E \cap K$, which is a group of order 2. Furthermore, $S = S'$ by Proposition 3.5, and therefore $S \cap K \leq \mathfrak{Z}S \cap S'$ is isomorphic to a subgroup of the Schur multiplier of $\bar{S}$. In particular, the Schur multiplier of $\bar{S}$ has even order. Hence by Schur [9, p. 108, Theorem V] the Sylow 2-subgroups of $\bar{S}$ are not quasidihedral groups. So they are dihedral, and we are finished by Proposition 3.6.

THEOREM C. *Assume that each involution in* S *acts trivially on* Ω *and that* $4 \leq |\Omega| \equiv 2 \pmod 4$. *Then* $S \cong SL(2, q)$ *and* $|\Omega| = q + 1$ *for a suitable odd prime power* q.

PROOF. Let (Ω, G) be a counterexample of smallest possible degree. By our assumption $|\Omega|$ is even so that because of Proposition 3.5 each element in S induces an even permutation on Ω. Hence each involution in $\bar{S}$ fixes at least 2 points. Suppose that each involution in $\bar{S}$ fixes exactly 2 points. Then all involutions in $\bar{S}$ are conjugate by Proposition 3.4(b), and therefore the Sylow 2-subgroups of $\bar{S}$ are cyclic or dihedral by Corollary 2.2. But then G is no counterexample because of Proposition 3.6. Hence $\bar{S}$ contains involutions fixing more than two points. Concerning such involutions we prove

(1) <u>If</u> u <u>is an involution in</u> $\bar{S}$ <u>fixing more than 2 points and</u> U <u>is the pre-image of</u> $\langle u \rangle$ <u>in</u> G, <u>then</u> $S(U) \cong SL(2, q)$ <u>and</u> $\overline{S(U)} \cong PSL(2, q)$ <u>for a suitable odd prime power</u> q. <u>Furthermore</u>, $|\Omega(U)| = q + 1 < (|\Omega| - 1)^{\frac{1}{2}} + 1$.

PROOF. As $u \in \bar{S}$, it is an even permutation, so that $|\Omega(U)| \equiv |\Omega| \equiv 2 \pmod 4$. Hence we can apply induction to the group space $(\Omega(U), \mathcal{L}_G U)$. It follows that $S(U) \cong SL(2, q)$ and $|\Omega(U)| = q + 1$ for a suitable odd prime power q. By our assumption the unique involution in S(U) must lie in K, so that $\overline{S(U)} \cong PSL(2, q)$. Because of Theorem A we have the inequality $|\Omega| > q^2 + 1$.

Suppose now that $\bar{S}$ contains an involution fixing exactly two points. By the above, $\bar{S}$ contains involutions which fix more than two points. Hence by Proposition 3.4(c) $\bar{S}$ contains an involution u such that $|\Omega(u)| \geq (|\Omega| - 1)^{\frac{1}{2}} + 1$. But this is a contradiction to (1). Hence each involution in $\bar{S}$ fixes at least three points in Ω.

(2) <u>If</u> α, β, <u>and</u> γ <u>are pairwise different points in</u> Ω, <u>then</u> $\bar{S}_{\alpha\beta\gamma}$ <u>does not contain any elementary abelian subgroup of order</u> 4.

PROOF. Suppose that $\bar{S}_{\alpha\beta\gamma}$ does contain an elementary abelian subgroup X of order 4. Let $X = \{1, a_1, a_2, a_3\}$ such that $|\Omega(a_1)| \geq |\Omega(a_2)| \geq |\Omega(a_3)|$. As we have seen, $|\Omega(a_3)| \geq 3$. Also a_2 acts on $\Omega(a_1)$. If this action is trivial, then $\Omega(a_1) = \Omega(a_2) = \Omega(a_3)$. But this implies that $\mathcal{L}_{\bar{N}_\alpha} a_1 = \mathcal{L}_{\bar{N}_\alpha} a_2 = \mathcal{L}_{\bar{N}_\alpha} a_3 = \mathcal{L}_{\bar{N}_\alpha} X < \bar{N}_\alpha$, contradicting the Brauer-Wielandt Lemma 2.4. So a_2 induces an involution on $\Omega(a_1)$. This involution fixes at least 3 points, namely α, β, and γ. Therefore $S(A_1) \cong SL(2, q^2)$, $|\Omega(a_1)| = q^2 + 1$, and $|\Omega(X)| = q + 1$, where q is a suitable prime power and A_1 is the pre-image of $\langle a_1\rangle$ in G. Using the Brauer-Wielandt Lemma we now obtain

$$(|\Omega| - 1)\cdot q^2 = |\bar{N}_\alpha||\mathcal{L}_{\bar{N}_\alpha} X|^2 = |\mathcal{L}_{\bar{N}_\alpha} a_1||\mathcal{L}_{\bar{N}_\alpha} a_2||\mathcal{L}_{\bar{N}_\alpha} a_3| \leq q^6$$

so that

$$|\Omega| \leq q^4 + 1.$$

But this again is impossible by (1).

Since $\bar{S}$ is doubly transitive, it contains an involution t. Let T be the pre-image of $\langle t\rangle$ in G. As we have seen, $|\Omega(T)| \geq 3$ and hence $\overline{S(T)} \cong PSL(2, q)$ for a suitable odd prime power q. We consider the group $\langle t\rangle \times \overline{S(T)}$. Let X be a subgroup of $\overline{S(T)}$ isomorphic to the alternating group A_4. (A subgroup of this kind exists. See [7, p.213,

Hauptsatz 8.27].) Let F be the subgroup of order 4 and x an element of order 3 in X, and denote $E = \langle t \rangle \times F$. Now x centralizes t and hence normalizes E. In fact, $\langle x \rangle$ has three orbits in $E - \{1\}$, namely x, $F - \{1\}$, and a second orbit of length 3.

We now choose an element $d \in F - \{1\}$ and denote the pre-image of $\langle d \rangle$ in G by D. We consider the centralizer $\mathcal{C}_{\bar{G}}\langle d \rangle$. In this centralizer we have the normal subgroup $\overline{S(D)}$ and the product $E \cdot \overline{S(D)}$. We denote the first group by $\hat{S}$ and the second by L. Note that $\hat{S} \cong PSL(2, \hat{q})$ for some odd prime power $\hat{q}$. Let α, β, and γ be three pairwise different points in $\Omega(D)$. Then $L_{\alpha\beta\gamma} \cap \hat{S} \quad 1$ and therefore

$$L_{\alpha\beta\gamma} \cong L_{\alpha\beta\gamma}\hat{S}/\hat{S} \leq L/\hat{S} = E\hat{S}/\hat{S} \cong E/E \cap \hat{S}.$$

Hence $L_{\alpha\beta\gamma}$ is an elementary abelian 2-group, and by (2) we obtain $|L_{\alpha\beta\gamma}| \leq 2$, so that

$$L_{\alpha\beta\gamma} = \langle d \rangle.$$

This implies that $|L| \leq 2(\hat{q} + 1)\hat{q}(\hat{q} - 1)$ and hence $|E \cap \hat{S}| \geq 2$, so that $|E \cap \hat{S}| = 2$ or 4.

We consider at first the case $|E \cap \hat{S}| = 4$. In this case the normalizer $\mathcal{N}_{\hat{S}}(E \cap \hat{S})$ contains an element y which induces an automorphism of order 3 of $E \cap \hat{S}$ (see Lemma 2.7). Here y centralizes $\langle d \rangle$ and hence normalizes $E = \langle d \rangle \times (E \cap \hat{S})$. Also, y has three orbits in $E - \{1\}$, namely $\{d\}$, $(E \cap \hat{S})^{\#}$, and a third orbit of length 3. As $|F \cap (E \cap \hat{S})| = 2$, $d^{\langle x,y \rangle} \geq F^{\#} \cup (E \cap \hat{S})^{\#}$ and actually $d^{\langle x,y \rangle} \geq E^{\#} - \{t\}$. Also, t is not invariant under y.

This implies that $\langle x, y\rangle$ is transitive on $E^{\#}$, which leads to a contradiction to Theorem 2.1.

Assume now that $|E \cap \hat{S}| = 2$. Then $L/\hat{S}$ is an elementary abelian group of order 4, and we have three subgroups H_1, H_2, and H_3 in L containing $\hat{S}$ as a subgroup of index 2. Let $H \in \{H_1, H_2, H_3\}$ and assume that $H_{\alpha\beta\gamma} \neq 1$. Then $H_{\alpha\beta\gamma} = L_{\alpha\beta\gamma} = \langle d\rangle$ and $H = \langle d\rangle \times \hat{S}$. So two of the groups H_1, H_2, and H_3 have a trivial stabilizer on α, β, and γ, which because of their order implies that they are sharply triply transitive on $\Omega(D)$. Choose $H \in \{H_1, H_2, H_3\}$ such that H is of this kind. Then we have $|E \cap H| = 4$ and $L = \langle d\rangle \times H$. Let Q be a Sylow 2-subgroup of H containing $E \cap H$. By Lemma 2.6, Q is a dihedral group or a quasidihedral group. As $|H| = (\hat{q} + 1)\hat{q}(\hat{q} - 1)$, we have $|Q| \geq 8$. Hence there exists a subgroup Y such that $(E \cap H) \triangleleft Y \leq Q$ and $|Y:(E \cap H)| = 2$. Let $z \in Y - (E \cap H)$. Then z does not centralize E, since in a dihedral or quasidihedral group each elementary abelian group of order 4 is self-centralizing. On the other hand, z centralizes d and hence normalizes $E = \langle d\rangle \times (E \cap H)$. So z induces a transvection in E. We consider the group $\langle x, z\rangle$ acting on the projective plane determined by E: If $\langle x, z\rangle$ has three orbits, then these orbits must be $\{t\}$, $F^{\#}$, and the remaining elements in $E^{\#}$. This implies that the elation z fixes t, d, and F so that $\langle t, d\rangle$ is its axis and d its centre. But then z fixes a line not passing through its centre, namely

$E \cap H$, a contradiction. So $\langle x, z \rangle$ has at most two orbits in $E^{\#}$, and we can again apply Theorem 2.1.

4. PROJECTIVE PLANES OF LENZ TYPE III

THEOREM 4.1. Let $(\mathcal{P}, \mathcal{L})$ be a projective plane, $\ell \in \mathcal{L}$, $P \in \mathcal{P} - \ell$ and Ω a subset of ℓ containing at least 3 points. Assume that $(\mathcal{P}, \mathcal{L})$ admits a finite group G of collineation leaving invariant P and Ω such that for all $X \in \Omega$ the group $G(X, XP)$ is transitive on $\Omega - \{X\}$. Assume in addition that each involution in G is a homology and that $|\Omega| \equiv 2 \pmod 4$. Then there exists an odd prime power q such that

$$\langle G(X, XP) \mid X \in \Omega \rangle \cong SL(2, q), \quad |\Omega| = q + 1,$$

and $(\mathcal{P}, \mathcal{L})$ contains a desarguesian subplane of order q.

PROOF. We consider the group space (Ω, G). Clearly this group space has the property (*) defined in section 3. If each involution in the subgroup $S = \langle G(X, XP) \mid X \in \Omega \rangle$ lies in $G(P, \ell)$, then the first two statements of our theorem follow from Theorem C. Assume that there exists an involution $t \in S - G(P, \ell)$. Then t must be a homology whose centre Z lies on ℓ and whose axis a passes through P, because obviously G leaves invariant the line ℓ. By Proposition 3.5 each element in S induces an even permutation of Ω. This implies that t must have fixed points in Ω, so that Z and $a \cap \ell$ lie in Ω. Also, (Ω, G)

is doubly transitive, and hence G contains an element x interchanging Z and $a \cap \ell$. But then $t^x \in G(a \cap \ell, ZP)$. By Lemma 2.9 the product tt^x is an involutory homology in $G(P, \ell)$. Also, this lemma implies that t, t^x, and tt^x are the only involutions in the stabilizer $G_{Z, a \cap \ell}$. Hence we have all properties required in Theorem B. So in both cases $\langle G(X, XP) | X \in \Omega \rangle \cong SL(2, q)$ and $|\Omega| = q + 1$ for a suitable prime power q. By [4, Theorem 2.8] we now obtain that $(\mathcal{P}, \mathcal{L})$ contains a subplane of order q.

THEOREM 4.2. _Let_ $(\mathcal{P}, \mathcal{L})$ _be a projective plane of finite odd order which contains a line_ ℓ _and a point_ P _not incident with_ ℓ _such that_ $(\mathcal{P}, \mathcal{L})$ _is_ (X, XP)-_transitive for each point_ $X \in \ell$. _Then_ $(\mathcal{P}, \mathcal{L})$ _is desarguesian_.

PROOF. Let $(\mathcal{P}, \mathcal{L})$ be a minimal counterexample and denote

$$G = \langle (X, XP) | X \in \ell \rangle.$$

We consider the group space (ℓ, G), which again has the property (*) of section 3. By Baer's Involution Lemma 2.8, each involution in G is a homology or a Baer involution. Suppose that G contains a Baer involution t. Then the set $\mathcal{Q}$ of fixed points of t generates a subplane of order q and $(\mathcal{P}, \mathcal{L})$ has order q^2 for some odd q. This subplane contains P and ℓ, and because of 3.1 it is again (X, XP)-transitive for each $X \in \ell \cap \mathcal{Q}$. Hence, by induction, $\mathcal{Q}$ is desarguesian. This implies that q is a prime power and S(t) induces a group of collineations of $\mathcal{Q}$ which is isomorphic to SL(2, q). (We define S(t) as in section 3.) Hence the group of permutations of $\mathcal{P} \cap \ell$ induced by S(t) is isomorphic to PSL(2, q), but this is impossible by Theorem A.

Thus each involution in G is a homology. If $|\ell| \equiv 2 \pmod 4$, then Theorem 4.1 implies that $(\mathcal{P}, \mathcal{L})$ is desarguesian. Assume that $|\ell| \equiv 0 \pmod 4$. Then each involution in G lies in G(P, ℓ), as otherwise it induces an odd permutation of ℓ. But $|G(P, \ell)| \mid |\ell| - 2 \equiv 2 \pmod 4$. This implies that the Sylow 2-subgroups of G contain only one involution. So they are cyclic or quaternion groups, and Proposition 3.6 implies that G $\cong$ SL(2, q) for a suitable prime power q. As in 4.1 this finishes our proof.

REFERENCES

1. R. Baer. Projectivities with fixed points on every line of the plane. Bull. Amer. Math. Soc. 52 (1946): 273-286.
2. D. Gorenstein and J.H. Walter. The characterization of finite groups with dihedral Sylow 2-subgroups. I.J. Algebra 2 (1965): 85-151.
3. C. Hering. Zweifach transitive Permutationsgruppen, in denen 2 die maximale Anzahl von Fixpunkten von Involutionen ist. Math. Z. 104 (1968): 150-174.
4. —— On projective planes of type VI. To appear.
5. C. Hering and W.M. Kantor. On the Lenz-Barlotti classification of projective planes. Arch. Math. 22 (1971): 221-224.
6. C. Hering, W.M. Kantor, and G.M. Seitz. Finite groups with a split BN-pair of rank 1. J. Algebra 20 (1972): 435-475.
7. B. Huppert. Endliche Gruppen I. Berlin-Heidelberg-New York: Springer Verlag (1967).
8. T.G. Ostrom. Double transitivity in finite projective planes. Canad. J. Math. 8 (1956): 563-567.
9. I. Schur. Untersuchungen über die Darstellung der endlichen Gruppen durch gebrochene lineare Substitutionen. J. Reine Angew. Math. 132 (1907): 85-137.

10. H. Wielandt. Beziehungen zwischen den Fixpunktzahlen von Automorphismengruppen einer endlichen Gruppe. Math. Z. 73 (1960): 146-158.
11. H. Zassenhaus. Kennzeichnung endlicher linearer Gruppen als Permutationsgruppen. Abh. Math. Sem. Univ. Hamburg 11 (1936): 17-40.

SOME RECENT RESULTS ON INCIDENCE GROUPS

H. Karzel

1. INTRODUCTION

For the study and foundation of absolute planes, it is useful to consider the associated kinematic space which is also called a group space. To each line X of an absolute plane there corresponds exactly one line reflection $\widetilde{X}$ fixing all points of X and preserving the structure of the absolute plane (compare [12], section 17). Let $\langle D \rangle$ be the set of all line reflections, $\Gamma := \langle D \rangle$ the group generated by all line reflections, $D^2 := \{XY: X, Y \in D\}$ (we denote the line and the line reflection by the same letter), and $J := \{\gamma \in \Gamma : \gamma^2 = 1, \gamma \neq 1\}$. If $\alpha \in \Gamma$, let $\overset{\frown}{\alpha} := \{X \in D: \alpha X \in J\}$; when $\alpha \in D^2 \backslash \{1\}$, the set $\overset{\frown}{\alpha}$ is called a <u>pencil</u>. Let L be the set of all pencils. Two pencils $\overset{\frown}{\alpha}, \overset{\frown}{\beta} \in L$ are called <u>joinable</u> if $\overset{\frown}{\alpha} \cap \overset{\frown}{\beta} \neq \phi$, and a pencil which is joinable to every other pencil is called a <u>proper pencil</u>. Let L_0 denote the set of all proper pencils. The pair Γ, D is then a <u>reflection group</u>; that is, Γ, D fulfills the following four axioms ([12], section 19):

(S1) D is a system of generators of Γ.

(S2) If $\widehat{\alpha} \in L$ and X, Y, Z $\in \widehat{\alpha}$, then XYZ $\in$ D.

(S3) $L_0 \neq \phi$.

(S4) For each X $\in$ D, the cardinality $|\{\widehat{\alpha} \in L_0: X \in \widehat{\alpha}\}| \neq 1, 2$.

If Γ is a reflection group, one can prove that D^2 is a subgroup of Γ and that, for each $\alpha \in D^2\backslash\{1\}$, the set $\widehat{\alpha}^2 := \{XY: X, Y \in \widehat{\alpha}\}$ forms a commutative subgroup of D^2. The kinematic space associated with a reflection group Γ, D is defined as follows: the point set of the kinematic space is the subgroup $G := D^2$ and the set of lines is $\mathcal{G} := \{\alpha\cdot\widehat{\beta}^2: \alpha, \beta \in D^2, \beta \neq 1\}$; this means that each line is a subgroup or a coset of D^2. One proves that the pair $(G, \mathcal{G})$ is an incidence space (see section 2). Therefore the set G is provided at the same time with a group structure and an incidence structure. Both structures fulfill the following compatibility conditions:

$\underline{V_\ell}$ For each $\gamma \in G$, the mapping $\gamma_\ell: \{\begin{matrix} G \to G \\ \xi \to \gamma\xi \end{matrix}$ is an automorphism of the geometric structure $(G, \mathcal{G})$.

$\underline{V_r}$ For each $\gamma \in G$, the mapping $\gamma_r: \{\begin{matrix} G \to G \\ \xi \to \xi\gamma \end{matrix}$ is an automorphism of the geometric structure $(G, \mathcal{G})$.

$\underline{V_k}$ Each line $X \in \mathcal{G}$ with $1 \in X$ is a subgroup of G.

REMARK. The kinematic space of an absolute plane is a (3-dimensional) projective space or a (3-1)-slit space (see section 2) if and only if the incidence structure of the absolute plane is a projective plane or an affine plane (see [3], [15] respectively); such an absolute plane is called a (generalized) elliptic or euclidean plane, respectively.

This situation led us to introduce the concept of an incidence group [1]:

A triple $(G, \mathcal{G}, \cdot)$ is called an incidence group if $(G, \mathcal{G})$ is an incidence space, $(G, \cdot)$ is a group, and the compatibility condition $\underline{V_\ell}$ is valid.

This paper will be a continuation of the reports [4] and [11] on incidence groups. In section 2, we give the geometric notations used and, in section 3, the algebraic ones. For the algebraic description of the class of punctured affine incidence groups, G. Kist has introduced the concept of a near-field extension, which is a generalization of the normal near-fields used for the algebraic representation of projective incidence groups (4.1). Many of the known infinite proper near-fields F do not contain sub-near-fields K such that the pair (F, K) is a normal near-field, but they do contain sub-near-fields such that they are near-field extensions. For instance, the Kalscheuer near-fields (section 3) are near-field extensions over $\mathbb{R}$ and over $\mathbb{C}$, but not normal.

Examples of incidence groups are obtained from algebraic structures (section 4), for instance, from associative unitary algebras. An important class consists of the kinematic spaces which are derivable from kinematic algebras. In section 5, we state the representation theorem for punctured affine incidence groups by near-field extensions of G. Kist.

A type of generalized incidence groups are the so-called 2-incidence groups (section 6), which can be described by pairs of near-domains. The case in which the underlying geometric structure is an affine space or plane will be investigated.

Finally a geometric characterization of projective kinematic spaces is briefly discussed in section 7.

2. GEOMETRIC CONCEPTS

Let P be a set and $\mathcal{G}$ a subset of the power set of P; the elements of P will be named points and the elements of $\mathcal{G}$ lines. The pair $(P, \mathcal{G})$ is called an incidence space if:

(I1) For each pair of distinct points $p, q \in P$, there exists exactly one line $A \in \mathcal{G}$ such that $p, q \in A$; this line will be designated by $\overline{p, q}$.

(I2) For each $X \in \mathcal{G}$, the cardinality $|X|$ is at least 2.

For the following, let $(P, \mathcal{G})$ be an incidence space. A subset $T \subset P$ is called a subspace if $t_1, t_2 \in T$, $t_1 \neq t_2$ implies $\overline{t_1, t_2} \subset T$. Let $\mathcal{T}$ be the set of all subspaces. Then $\mathcal{T}$ is closed with regard to intersections and therefore the closure of a subset $M \subset P$ defined by $\bar{M} := \cap\{T:$

$M \subset T \in \mathcal{T}\}$ is a subspace. For $T \in \mathcal{T}$ we define by $\dim T := \inf\{|M| : \bar{M} = T\} - 1$ the dimension of a subspace. Subspaces of dimension 2 are called planes. For a point $p \in P$ let $\mathcal{G}(p) := \{X \in \mathcal{G} : p \in X\}$ and for a subspace $T \in \mathcal{T}$ let $\mathcal{G}(T) := \{X \in \mathcal{G}: X \subset T\}$. Let $m \in \mathbb{N} \cup \{0\}$. A line $L \in \mathcal{G}$ is called an m-line if: (1) $|L| \geq 3$ if $m = 0$, and (2) for each $X \in P\backslash L$, there exist exactly m distinct lines $H_1, \ldots, H_m \in \mathcal{G}(x) \cap \mathcal{G}(\overline{\{x\} \cup L})$ such that $H_i \cap L = \phi$ for $i \in \{1, 2, \ldots, m\}$ and each $X \in \mathcal{G}(\overline{\{x\} \cup L})$ with $|X \cap H_i| = 1$ for all $i \in \{1, \ldots, m\}$ satisfies $X \cap L \neq \phi$. Let $\mathcal{G}_m$ be the set of all m-lines.

REMARKS. An incidence space $(P, \mathcal{G})$ with $\mathcal{G} = \mathcal{G}_0$ is a projective space. If $\mathcal{G} = \mathcal{G}_1$ we call $(P, \mathcal{G})$ a pseudoaffine space; if in addition the relation on $\mathcal{G}$ defined by "$A \parallel B \Longleftrightarrow A = B$ or $A \cap B = \phi$ and $\dim \overline{A \cup B} = 2$" is transitive, then $(P, \mathcal{G})$ is an affine space. This last condition can be proved if there is a line $A \in \mathcal{G}$ with $|A| \geq 4$. (Compare [11], pp. 79-81).

For every subset $U \subset P$ we define the trace space $(U, \mathcal{G}_u)$ by $\mathcal{G}_u := \{X \cap U: X \in \mathcal{G}, |X \cap U| \geq 2\}$. The trace space is again an incidence space. If $(P, \mathcal{G})$ is a projective space and L a subspace of $(P, \mathcal{G})$, then the trace space $(P\backslash L, \mathcal{G}_{(P\backslash L)})$ is called a slit space and if $\dim P = n$ and $\dim L = r$ it is called an $(n - r)$-slit space. An internal characterization of slit spaces is given in

[11]. If $(P, \mathcal{G})$ is a pseudoaffine (affine) space and q a fixed point of P, then the trace space $(P\backslash\{q\}, \mathcal{G}_{(P\backslash\{q\})})$ is called a <u>punctured pseudoaffine</u> (<u>punctured affine</u>) <u>space</u>. Each line of $(P\backslash\{q\}, \mathcal{G}_{(P\backslash\{q\})})$ is either a 1-line or a 2-line. G. Kist [13] has given the following internal characterization : an incidence space $(P, \mathcal{G})$ is a punctured pseudoaffine space if $\mathcal{G} = \mathcal{G}_1 \cup \mathcal{G}_2$ and if the conditions (P1) and (P2) hold, and a punctured affine space if (P3) also holds:

(P1) Let E be a plane with $\mathcal{G}_2(E) := \mathcal{G}(E) \cap \mathcal{G}_2 \neq \phi$; then for each point $x \in E$, $|\mathcal{G}(x) \cap \mathcal{G}_2(E)| = 1$.

(P2) To every line $A \in \mathcal{G}_1$ there exists a line $B \in \mathcal{G}_2$ such that $A \parallel B$ (that is, $A \cap B = \phi$ and $\dim \overline{A \cup B} = 2$).

(P3) The relation $\parallel_1 := \parallel \cap (\mathcal{G}_1 \times \mathcal{G}_1)$ is transitive or there is a line $A \in \mathcal{G}$ with $|A| \geq 4$.

3. ALGEBRAIC CONCEPTS (NEAR-DOMAINS, NEAR-FIELDS, NEAR-FIELD EXTENSIONS)

A set $(F, +, \cdot)$ provided with two operations $+: F \times F \to F$ and $\cdot: F \times F \to F$ is called a <u>near-domain</u> [5], p. 123) if the following conditions hold:

(F1) $(F, +)$ is a loop (0 denotes the neutral element) such that $a, b \in F$ and $a + b = 0$ implies $b + a = 0$.

(F2) $(F^*, \cdot)$ with $F^* := F\backslash\{0\}$ is a group.

(F3) For all a, b, c $\in$ F, $a\cdot(b + c) = a\cdot b + a\cdot c$.

(F4) For every a $\in$ F, $0\cdot a = 0$.

(F5) For all a, b $\in$ F* there exists $d_{a,b} \in$ F such that for all x $\in$ F, $a + (b + x) = (a + b) + d_{a,b}\cdot x$.

A near-domain (F, +, $\cdot$) is called a <u>near-field</u> if (F, +) is a group. A set (F, +, $\circ$, $\cdot$) provided with three binary operations is called a <u>dicksonian near-field</u> (or regular near-field) [2] if (1) (F, +, $\circ$) is a near-field, (2) (F, +, $\cdot$) is a field, and (3) for all a $\in$ F* the mapping ψ_a: F $\to$ F; $x \to a^{-1}\cdot(a \circ x)$ (a^{-1} denotes the inverse with respect to $\cdot$) is an automorphism of the field (F, +, $\cdot$).

REMARKS. (1) We know that every finite near-domain is a near-field but we do not know at present whether there are near-domains which are not near-fields. (2) With the exception of seven finite near-fields, all known near-fields are dicksonian near-fields.

A pair (F, K) is called a <u>normal near-field</u> if F is a near-field, K a sub-near-field $\neq$ F, and K* $\triangleleft$ F* [1]. As H. Wähling [17] proved, K is even a field and (F, K) a left vector space.

Normal near-fields are used for the algebraic representation of desarguesian projective incidence groups [1], [4]. For the description of desarguesian punctured affine incidence groups, G. Kist [13] introduces the concept of a near-field extension:

A quadruple $((F, +, \circ), (K, +, \circ), \cdot, \psi)$ consisting of a near-field $(F, +, \circ)$ a sub-near-field $(K, +, \circ)$ of $(F, +, \circ)$, a mapping $\cdot : F \times K \to F$, and a mapping $\psi : F^* \to S_k; a \to \psi_a$ in the symmetric group of the set K is called a near-field extension if the following conditions are valid:

(E1) $\cdot(K \times K) =: K \cdot K \subset K$ and $(K, +, \cdot)$ is a field such that the identity $1_\circ$ of $(K, \circ)$ and $1.$ of $(K, \cdot)$ coincide.

(E2) $((F, +), (K, +, \cdot), \cdot)$ is a right vector space.

(E3) For all $a, b \in F^*$, and for all $\lambda \in K$ holds $a \circ (b \cdot \lambda) = (a \circ b) \cdot \psi_a(\lambda)$.

In [13] Kist proves the properties (3.1) of near-field extensions and shows how one can get examples of near-field extensions.

(3.1) Let $(F, K, \cdot, \psi)$ be a near-field extension, then:

(a) For every $a \in F^*$ the mapping ψ_a is an automorphism of $(K, +, \cdot)$.

(b) $(K, +, \circ, \cdot)$ is a dicksonian near-field.

(3.2) Let $(F, +, \circ)$ be a near-field, $K_F := \{x \in F: \text{for every } y, z \in F, (y + z)x = yx + zx\}$ the kernel (the kernel of a near-field is always a field!), $\cdot := \circ|F \times K_F$, and ψ the function mapping each $a \in F^*$ on the identity of S_{K_F}. Then $(F, K_F, \cdot, \psi)$ is a near-field extension. The same is true if H is a subfield of K_F.

REMARK. With (3.2) we can get examples of near-field extensions from the Kalscheuer near-fields: Let $(\mathbb{H},+,\cdot)$ be the quaternions over the reals $\mathbb{R}$, $c \in \mathbb{R}^*$,

$$\psi: \begin{cases} H^* \to S_{\mathbb{H}} \\ a \to \psi_a: \begin{cases} \mathbb{H} \to \mathbb{H} \\ x \to e^{-ic\log(a\cdot\bar{a})}\cdot x\cdot e^{ic\log(a\cdot\bar{a})} \end{cases} \end{cases}$$

where $\bar{}$ denotes the involutorial antiautomorphism

$$-: \begin{cases} \mathbb{H} \to \mathbb{H} \\ a = a_0+ia_1+ja_2+ka_3 \to \bar{a} = a_0-ia_1-ja_2-ka_3 \end{cases}$$

of $(\mathbb{H}, +, \cdot)$ ($a\cdot\bar{a}$ is an element of $\mathbb{R}^*_+ := \{r \in \mathbb{R}: 0 < r\}$) and

$$\circ: \begin{cases} \mathbb{H} \times \mathbb{H} \to \mathbb{H} \\ (a, b) \to \begin{cases} a\circ b := a\cdot\psi_a(b) & \text{if } a \neq 0, \\ 0 & \text{if } a = 0. \end{cases} \end{cases}$$

Then $(\mathbb{H}, +, \circ)$ is a Kalscheuer near-field. The kernel of $(\mathbb{H}, +, \circ)$ is the field of complex numbers $(\mathbb{C}, +, \cdot)$ (in $\mathbb{C} := \mathbb{R} + i\mathbb{R}$ the multiplications $\circ$ and $\cdot$ coincide). Therefore $((\mathbb{H}, +, \circ), (\mathbb{C}, +, \circ), \circ, \psi')$ and $((\mathbb{H}, +, \circ), (\mathbb{R}, +, \circ), \circ, \psi'')$ with $\psi'_a := \psi_a|\mathbb{C}\,(=id_{\mathbb{C}})$ and $\psi''_a := \psi_a|\mathbb{R}\,(=id_{\mathbb{R}})$ are near-field extensions.

(3.3) Let $((F, +, \circ), (K, +, \circ))$ be a normal near-field,

$$\cdot: \begin{cases} F \times K \to F \\ (a, \lambda) \to a\cdot\lambda := \lambda\circ a \end{cases}$$

and

$$\psi : \begin{cases} F^* \to S_K \\ a \to \psi_a : \begin{cases} K \to K \\ \lambda \to \lambda^a = a \circ \lambda \circ a^{-1}. \end{cases} \end{cases}$$

Then $((F, +, \circ), (K, +, \circ), \cdot, \psi)$ is a near-field extension and $((K, +, \cdot)$ is an anti-isomorphic field to $(K, +, \circ))$.

REMARK. (3.3) shows that every normal near-field can be considered as a near-field extension. However there are near-field extensions which are not normal near-fields. For instance, the Kalscheuer near-fields do not contain any sub-near-field K such that $((\mathbb{H}, +, \circ), (K, +, \circ))$ is a normal near-field.

(3.4) Let $(F, +, \circ, \cdot)$ be a dicksonian near-field and $(K, +, \circ)$ be a sub-near-field such that $a^{-1} \cdot (a \circ \lambda) \in K$ for all $a \in F^*$ and all $\lambda \in K$. Then $((F, +, \circ), (K, +, \circ), \odot, \psi)$ with $\odot = \cdot|F \times K$ and $\psi_a : \begin{cases} K \to K \\ \lambda \to a^{-1} \cdot (a \circ \lambda) \end{cases}$ is a near-field extension.

4. INCIDENCE GROUPS

An incidence group $(G, \mathcal{G}, \cdot)$ is called 2-sided if both V_ℓ and V_r are valid, and linearly fibred or a kinematic space if V_ℓ, V_r, and V_k are valid. If the incidence structure $(G, \mathcal{G})$ is for instance a projective, affine, or slit space we call $(P, \mathcal{G}, \cdot)$ a projective, affine, or slit incidence group.

Examples of incidence groups

1. Let (V, K) be a left vector space and $G := \{a + Kb: a, b \in V, b \neq 0\}$; then the pair (V, G) is an affine space and the pair (V, +) where + is the vector addition is a commutative group. Since for each $a \in V$ the mapping $a_\ell: V \to V; x \to a + x$ is bijective and maps lines on lines, (V, G, +) is a commutative affine incidence group.

2. Let (A, K) be an associative algebra over the field K, such that K is in the centre of A and let U be the set of all units of A. Now we can apply an affine and a projective derivation:

(a) By the affine derivation, let $G := \{a + bK: a, b \in A, b \neq 0\}$ and (U, G_u) be the trace space (section 2) belonging to the set of units. Then (U, G_u, $\cdot$) is a 2-sided incidence group, since U is a group with respect to the multiplication given in the algebra A. In the special case that (A, K) is a division algebra, we have $U = A^* := A\backslash\{0\}$ and (A^*, G_A, $\cdot$) is a punctured affine incidence group (as they will be discussed in section 5).

(b) For the projective derivation, if $A^*/K^* := \{K^*\cdot x: x \in A^*\}$ where $A^* := A\backslash\{0\}$ and $K^* := K\backslash\{0\}$, let L_2 be the set of all 2-dimensional vector subspaces of (A, K) and

$$\chi : \begin{cases} A^* \to A^*/K^* \\ x \to K^*x \end{cases} \quad \text{the canonical map.}$$

Then (A^*/K^*, G) with $G := \{\chi(L^*): L \in L_2\}$ is the projective space corresponding to the vector space (A, K) and since $K^* \lhd U$ the set $\chi(U) = U/K^*$ is a group. If we again take the trace space as incidence space, then $(U/K^*, G_{U/K^*}, \cdot)$ is a 2-sided incidence group. In the case that (A, K) is a division algebra, $(A^*/K^*, G, \cdot)$ is a projective incidence group. We obtain examples of kinematic spaces from kinematic algebras [7]: (A, K) is a kinematic algebra if for each $x \in A$ the element x^2 belongs to the set $K + Kx$. The quaternions $\mathbb{H}$ over the reals $\mathbb{R}$ constitute a kinematic division algebra and the corresponding kinematic space is the same, which is associated with the classical elliptic plane. The kinematic space of the classical euclidean plane can be derived from the four-dimensional algebra $(A, \mathbb{R})$ with basis $\{1, i, \varepsilon, i\varepsilon\}$ and $i^2 = -1$, $\varepsilon^2 = 0$, $i\varepsilon = -\varepsilon i$.

The projective incidence groups are the most intensely studied ones [4]. The main theorems on projective incidence groups are the following:

(4.1) (a) Let (F, K) be a normal near-field (section 3), $G := F^*/K^*$ and $G := \{\chi(L^*) : L \in L_2\}$, then $\Pi(F, K) := (G, G, \cdot)$ is a desarguesian projective incidence group [1, 4].

(b) Let $(G, G, \cdot)$ be a desarguesian projective incidence group, then there is exactly one normal near-field (F, K) such that $(G, G, \cdot)$ and $\Pi(F, K)$ are isomorphic [1, 4].

(4.2) Let $(G, \mathcal{G}, \cdot)$ be a desarguesian projective incidence group and (F, K) the corresponding normal near-field; then

(a) $(G, \mathcal{G}, \cdot)$ is 2-sided if and only if F is a field [16].

(b) $(G, \mathcal{G}, \cdot)$ is a commutative incidence group if and only if F is a commutative field [4].

(c) $(G, \mathcal{G}, \cdot)$ is a kinematic space if and only if (F, K) is a kinematic division algebra. If G is not commutative, then (F, K) is a quaternion division algebra (this means that K is the centre of F and $[F: K] = 4$). If G is commutative, then K is a commutative field of characteristic 2 and F a pure inseparable extension of K [6].

Similar results on slit incidence groups can be found in [11].

5. PUNCTURED AFFINE INCIDENCE GROUPS

The punctured affine incidence groups $(G, \mathcal{G}, \cdot)$ - here $(G, \mathcal{G})$ is a punctured affine space (section 2) - were recently studied by G. Kist in his dissertation [13]. He got the following main results:

(5.1) (a) Let $((F, +, \circ), (K, +, \circ), \circ, \psi)$ be a near-field extension (section 3) with $K \neq \mathbb{Z}_2$, $F \neq K$, and $\mathcal{G} := \{a + b\cdot K\colon a, b \in F^*, a \notin b\cdot K\} \cup \{bK^*\colon b \in F^*\}$; then $(F^*, \mathcal{G}, \circ)$ is a desarguesian punctured affine incidence group with $\mathcal{G}_1 = \{a + b\cdot K\colon a, b \in F^*, a \notin b\cdot K\}$ and $\mathcal{G}_2 =$

$\{b \cdot K^*: b \in F^*\}$.

(b) Let $(G, \mathcal{G}, \circ)$ be a desarguesian punctured affine incidence group; then there exists exactly one near-field extension $(F, K, \cdot, \psi)$ such that $(G, \mathcal{G}, \circ)$ and the incidence group derived from the near-field extension are isomorphic and the groups $(F^*, \circ)$ and $(G, \circ)$ coincide.

REMARK. In section 3 we have seen that every normal near-field can be considered as a near-field extension and that there are near-field extensions which do not come from normal near-fields. Those punctured affine incidence groups which can be represented by normal near-fields or even by field extensions are characterized by G. Kist [13].

6. 2-INCIDENCE GROUPS

In this section we shall consider the following generalization of incidence groups, which are closely related to the punctured affine incidence groups:

A tripel $(P, \mathcal{G}, \Gamma)$ is called a 2-incidence group if

(1) $(P, \mathcal{G})$ is an incidence space.

(2) (P, Γ) is a sharply 2-transitive group (this means that, for each two pairs (a_1, a_2), $(b_1, b_2) \in P \times P$ with $a_1 \neq a_2$, $b_1 \neq b_2$, there is exactly one $\gamma \in \Gamma$ such that $\gamma(a_1) = b_1$, $\gamma(a_2) = b_2$.

(3) Each $\gamma \in \Gamma$ is an automorphism of the incidence space $(P, \mathcal{G})$.

In this section, we say that two lines $X, Y \in \mathcal{G}$ are parallel, denoted by $X \parallel Y$, if $X = Y$ or if $\dim \overline{X \cup Y} = 2$ and $X \cap Y = \phi$. An automorphism α of $(P, \mathcal{G})$ will be called a dilatation if, for all $X \in \mathcal{G}$, $\alpha(X) \parallel X$ and a translation if $\alpha = 1$ or if α is a dilatation without fixed points.

To every 2-incidence group, there corresponds a usual incidence group:

(6.1) Let $(P, \mathcal{G}, \Gamma)$ be a 2-incidence group, $a, e \in P$ with $a \neq e$, $P_a := P \backslash \{a\}$, $\mathcal{G}_a := \mathcal{G}_{P_a}$ (:= the trace space $\{X \cap P_a : X \in \mathcal{G}, |X \cap P_a| \geq 2\}$), and $\Gamma_a := \{\gamma \in \Gamma : \gamma(a) = a\}$. Then Γ_a operates regularly on P_a; hence, for each $x \in P_a$ there is exactly one $\dot{x} \in \Gamma_a$ such that $\dot{x}(e) = x$. The multiplication $\cdot$ defined by

$$\cdot : \begin{cases} P_a \times P_a \to P_a \\ (x, y) \to \dot{x}(y) \end{cases}$$

makes P_a a group, such that $(P_a, \mathcal{G}_a, \cdot)$ is an incidence group.

(6.2) Each desarguesian punctured affine incidence group $(G, \mathcal{G}, \cdot)$ can be extended to a 2-incidence group $(P, \overline{\mathcal{G}}, \Gamma)$ such that for any $a, e \in P$, $a \neq e$, the incidence group $(P_a, \overline{\mathcal{G}}_a, \cdot)$ and $(G, \mathcal{G}, \cdot)$ are isomorphic. Let $((F, +, \circ), (K, +, \circ), \cdot, \psi)$ be the near-field extension corresponding to $(G, \mathcal{G}, \cdot)$, $\overline{\mathcal{G}} := \{a + b \circ K : a, b \in F, b \neq 0\}$ and $\Gamma := \{[a, b] : a, b \in F, b \neq 0\}$ the sharply 2-transitive group consisting of all mappings

$$[a, b] : \begin{cases} F \to F \\ x \to a + b \circ x, \end{cases}$$

then $(F, \bar{G}, \Gamma)$ _is a 2-incidence group extending_ $(G, G, \cdot) = (F^*, G, \circ)$.

For the proof, it is sufficient to remark that condition (E3) of a near-field extension implies that $b \circ K = b \cdot K$.

(6.3) _Let_ (P, G, Γ) _be a 2-incidence group_, $K \in G$ _and_ $\Gamma_K := \{\gamma \in \Gamma : \gamma(K) = K\}$. _Then_ (K, Γ_K) _is a sharply 2-transitive group_.

PROOF. Let $(a_1, a_2), (b_1, b_2) \in K \times K$ with $a_1 \neq a_2$, $b_1 \neq b_2$, and $\gamma \in \Gamma$ uniquely determined by $\gamma(a_1) = b_1$, $\gamma(a_2) = b_2$. Then $K = \overline{a_1, a_2} = \overline{b_1, b_2}$ and $\gamma(K) = \overline{\gamma(a_1), \gamma(a_2)} = \overline{b_1, b_2} = K$, so $\gamma \in \Gamma_K$.

For the further discussion of 2-incidence groups, we need the following theorems - from the theory of sharply 2-transitive groups stated for instance in section 11 of [5]:

(6.4) _Let_ $(F, +, \cdot)$ _be a near-domain; for each pair_ $(a, b) \in F \times F^*$, _let_ $[a, b]$ _be the mapping_ $[a, b]: F \to F;\ x \to a + b \cdot x$ _and_ $\Gamma := \{[a, b]: a, b \in F, b \neq 0\}$. _Then_ (F, Γ) _is a sharply 2-transitive group_.

(6.5) _Let_ (P, Γ) _be a sharply 2-transitive group_, $J :=$

$\{\gamma \in \Gamma : \gamma^2 = 1, \gamma \neq 1\}$ and $0, 1 \in P$ two distinct points, then:

(a) For all $p, q \in P$ with $p \neq q$, there is a unique $\alpha \in J$ such that $\alpha(p) = q$ and $\alpha(q) = p$.

(b) Let $\alpha, \beta \in J$, $\alpha \neq \beta$; then $\alpha\beta$ is free of fixed points.

(c) If one $\alpha \in J$ has a fixed point, then each $\iota \in J$ has a fixed point and to each point $x \in P$, there corresponds exactly one $\hat{x} \in J$ fixing x.

(d) There exist two operations $+: P \times P \to P$ and $\cdot: P \times P \to P$ such that $(P, +, \cdot)$ is a near-domain (section 3); 0 (resp. 1) is the neutral element with regard to + (resp. $\cdot$); Γ consists of all mappings $[a, b]: P \to P;\ x \to a + b\cdot x$ with $a, b \in P$ and $b \neq 0$ and $\Gamma_0 \cong (P^*, \cdot)$. For each $x \in P^* := P\backslash\{0\}$, let $\overset{+}{x}$ be the involution interchanging 0 and x; then + is given by

$$a + b := \begin{cases} \overset{+}{a}\hat{0}(b) & \text{if } a \neq 0 \\ b & \text{if } a = 0 \end{cases}$$

or

$$a + b := \begin{cases} \overset{+}{a}(b) & \text{if } a \neq 0 \\ b & \text{if } a = 0 \end{cases}$$

according as the involutions of J have fixed points or not.

(e) The near-domain $(P, +, \cdot)$ has the characteristic 2 (that means $1 + 1 = 0$) if and only if the elements of J have no fixed points.

By Theorem (6.5) the sharply 2-transitive groups (P, Γ) decompose into two classes depending on whether each involution $\iota \in J$ has a fixed point or not. A 2-incidence group (P, G, Γ) is said to be of type 1 if each $\iota \in J$ has a fixed point; otherwise it is of type 2.

From (6.3) and (6.5) we obtain:

(6.6) *Let* (P, G, Γ) *be a 2-incidence group*, $0, 1 \in P$ *two distinct points*, $K := \overline{0, 1}$, *and* $(P, +, \cdot)$ *the near-domain associated with* (P, Γ) *whose neutral elements are* 0 *and* 1. *Then* Γ *consists of all mappings* $[a, b] : P \to P;\ x \to a + b \cdot x$ *with* $a, b \in P$, $b \neq 0$, K *is a sub-near-domain, and each line* $X \in G$ *can be represented in the form* $X = a + b \cdot K$ *with* $a, b \in P$ *and* $b \neq 0$.

The last statement follows from the fact that Γ is 2-transitive on P and, hence, it is transitive on the set G of all lines.

Conversely we have:

(6.7) *Let* $(F, +, \cdot)$ *be a near-domain*, $(K, +, \cdot)$ *a proper sub-near-domain of* $(F, +, \cdot)$, (F, Γ) *the corresponding sharply 2-transitive group* (see (6.4)) *and* $G := \{a + b \cdot K : a, b \in F, b \neq 0\}$. *Then* (F, G, Γ) *is a 2-incidence group*.

PROOF. Let $X := a + b \cdot K \in G$ and $[u, v] \in \Gamma$, then $[u, v](X) = u + v(a + b \cdot K) = u + (v \cdot a + v \cdot b \cdot K) = (u + v \cdot a) + d_{u,v \cdot a} \cdot v \cdot b \cdot K \in G$ (see section 3); this shows that Γ preserves the structure $(F, \mathcal{G})$. Let $a, b \in F$ be two distinct points. By section 3 we obtain $a + (-a + b)\cdot 1 = a + (-a + b) = 0 + d_{a,-a}\cdot b = b$. Therefore $X := a + (-a + b)\cdot K$ is a line joining a and b. By the sharp 2-transitivity of Γ, it is enough to show that K is the unique line joining the points 0 and 1. Let $0, 1 \in a + b\cdot K$; then there is $\lambda \in K$ such that $0 = a + b\lambda$. Thus $a = -b\cdot\lambda = b(-\lambda)$ and $a + b\cdot K = b(-\lambda) + b\cdot K = b\cdot(-\lambda + K) = b\cdot K$. Since $1 \in a + b\cdot K = b\cdot K$ implies the existence of $\mu \in K$ with $1 = b\cdot\mu$, we see that $b = \mu^{-1} \in K$ and therefore $a + b\cdot K = b\cdot K = \mu^{-1}\cdot K = K$. Each line $a + b\cdot K \in \mathcal{G}$ contains (at least) the points a and $a + b$. Therefore $(F, \mathcal{G})$ is an incidence space and $(F, \mathcal{G}, \Gamma)$ a 2-incidence group.

By the Theorems (6.6) and (6.7) we see that there is bijective correspondence between all 2-incidence groups $(P, \mathcal{G}, \Gamma)$ and all pairs $((F, +, \cdot), (K, +, \cdot))$ of near-domains, where K is a sub-near-domain of F. We shall now study this correspondence for the class of 2-incidence groups, where $(P, \mathcal{G})$ is an affine space.

First we need the following lemma:

(6.8) <u>Let</u> $(P, \mathcal{G}, \Gamma)$ <u>be a 2-incidence group of type</u> 1. <u>Then</u>:

(a) <u>For all</u> $x \in P$ <u>and all</u> $X \in \mathcal{G}$ <u>with</u> $x \in X$, $\hat{x}(X) = X$.

(b) <u>For all</u> $x \in P$ <u>and all</u> $Y \in \mathcal{G}$ <u>with</u> $x \notin Y$, $\hat{x}(Y) \cap Y = \phi$.

(c) *For all* $x \in P$, *the involution* $\hat{x}$ *is a dilatation*.

PROOF. (a) From (6.3), (6.5), and (6.6) it follows that (X, Γ_X) is a sharply 2-transitive group of type 1. Therefore $J_X = \{\hat{y}: y \in X\}$ (see (6.5c)) is the set of involutions of Γ_X and $\hat{x}(X) = X$ as required.

(b) Suppose, if possible, that $z \in Y \cap \hat{x}(Y)$; then $z \neq x$, $\hat{x}(\overline{x, z}) = \overline{x, z}$ by (a), and $\overline{x, z} \cap Y = \overline{x, z} \cap \hat{x}(Y) = \{z\}$ so $\hat{x}(z) = z$. Since $\hat{x}$ thus has the two distinct fixed points x and z, the sharp 2-transitivity implies that $\hat{x} = 1$, contradicting $\hat{x} \in J$.

(c) The statement (c) is a consequence of (a) and (b), since by (a) any plane E containing x is fixed by $\hat{x}$.

(6.9) *Let* (P, G, Γ) *be an affine 2-incidence group with* $|X| \geq 3$ *for* $X \in G$ *and* $\dim P \geq 2$. *Suppose* $((P = F, +, \circ), (K, +, \circ))$ *are an associated pair of near-domains and*

$$T := J \cdot J = \{\hat{x}\hat{y}: x, y \in P\}$$

or

$$T := J \cup \{1\}$$

according as (P, G, Γ) *is of type* 1 *or of type* 2. *Then*:

(a) T *is a group consisting of translations which operates regularly on the set of points and* T *is isomorphic to* $(F, +)$, *if* (P, G, Γ) *is either of type* 1 *or of type* 2 *and* $\dim P = 2$.

(b) $(F, +, \circ)$ *is a near-field and* $(K, +, \circ)$ *a sub-near-field, if* T *is a group*.

(c) _If_ $\dim P = 2$, _then_ (P, G) _is a translation plane and_ $[F: K] = 2$ (_that is_, $F = K + aK$ _for_ $a \in F \setminus K$).

(d) _If_ (P, G) _is desarguesian and_ T _a group_, _then there exist two mappings_ $\cdot: F \times K \to F$ _and_ $\psi: F^* \to S_K$ _such that_ $((F, +, \circ), (K, +, \circ), \cdot, \psi)$ _is a near-field extension._

We prove first the statements (a), (b), (c) for 2-incidence groups (P, G, Γ) of type 1.

Results (6.8c) and (6.5c) imply that T is a set consisting of translations. Let $a, b \in P$ and $\gamma \in J$ be the involution interchanging a and b, so $\gamma\hat{a} \in T$ and $\gamma\hat{a}(a) = b$. Since by assumption (P, G) is an affine space of $\dim P \geq 2$, this implies that T is a commutative group and (P, G) a translation space. From (6.5d), we obtain $T \cong (P = F, +)$ and therefore $(F, +, \circ)$ is a near-field.

Now let (P, G, Γ) be of type 2 and $\dim P = 2$. Let $\iota \in J$ and $X \in G$ with $\iota(X) \neq X$. Then $\{a\} = X \cap \iota(X)$ would imply $\{\iota(a)\} = \iota(X) \cap X = \{a\}$ contradicting the assumption, that ι has no fixed points. Therefore and because $\dim P = 2$ we have $\iota(X) \| X$ for all $X \in G$. Hence ι is a translation.

Since any two distinct points can be interchanged by an involution, (P, G) is a translation plane and $T = J \cup \{1\}$ is the group of all translations. From (6.5d) we derive $T \cong (F, +)$ and $(F, +, \circ)$ is a near-field.

It remains to prove statement (d) for both types. By Theorems (6.1) and (6.5), the incidence group $(F^*, G_0, \circ)$, with $F^* = F_0 = F\setminus\{0\}$ and $G_0 := \{X\setminus\{0\}: X \in G\}$ corresponding to the points $0, 1 \in F$, is a desarguesian punctured affine incidence group and $(F^*, \circ) \cong \Gamma_0$. Applying the theorem (5.1) of Kist there is a near-field extension $((F, \oplus, \odot), (K, +, \odot), \cdot, \psi)$ representing $(F^*, G_0, \circ)$ such that $\odot = \circ$. Statement (a) finally implies that $\oplus = +$.

Problem: Let (P, G, Γ) be an affine 2-incidence group of type 2 with $\dim P \geq 3$ and $((F, +, \circ), (K, +, \circ))$ the associated pair of near-domains. Is $(F, +, \circ)$ a near-field and $(K, +, \circ)$ a sub-near-field? It can be proved that $(K, +, \circ)$ is a near-field and that there is an addition $\oplus: F \times F \to F$, such that $(F, \oplus, \circ)$ is a nearfield and $\oplus|K \times K = +|K \times K$. The problem is equivalent to the question whether $+ = \oplus$.

REMARK. Theorems (6.2) and (6.9) show that every desarguesian affine 2-incidence group can be obtained by extending a desarguesian punctured affine incidence group.

(6.10) *Let* $((F^*, +, \circ), (K, +, \circ), \cdot, \psi)$ *be a near-field extension with* $K \neq F$ *and* (F, G, Γ) *the 2-incidence group corresponding to the pair of near-domains* $((F, +, \circ), (K, +, \circ))$. *Then* (F, G) *is a desarguesian affine space*.

(6.11) *Let* $((F, +, \cdot), (K, +, \cdot))$ *be a pair of near-fields, such that* $(K, +, \cdot)$ *is a sub-near-field of* $(F, +, \cdot)$ *with*

$[F: K] = 2$, <u>and let</u> (F, G, Γ) <u>be the corresponding 2-incidence group</u>. <u>Then</u> (F, G) <u>is a translation plane</u>.

PROOF. We have to establish the parallel axiom. Since Γ operates transitively on G it is sufficient to consider the line K and a point $a \in P\backslash K$. Now $G(a) = \{a + bK: b \in F^*\}$ and if $b \in F\backslash K$, the equation $\lambda + b\mu = a$ has solutions λ, $\mu \in K$, because $F = K + bK$. This implies $\lambda = a + b(-\mu) \in K \cap a + bK$. Therefore $a + K$ is the unique line through a which does not intersect K. (F, G) is a translation plane, because all mappings $[a, 1]: x \to a + x$ for $a \in F$ are translations.

7. GEOMETRIC CHARACTERIZATIONS OF SOME CLASSES OF INCIDENCE GROUPS

In an incidence group $(G, G, \cdot)$ we call a subgroup H of $(G, \cdot)$ an <u>incidence subgroup</u> if H is a subspace of (G, G). For each incidence subgroup H of $(G, G, \cdot)$ the set $S := \{xH: x \in G\}$ of all left cosets of H in G forms a decomposition of G into subspaces which all have the same dimension. In particular if $(G, G, \cdot)$ is a projective incidence group, S is a spread - more precisely, a t-spread if $\dim H = t$. If (F, K) is the normal near-field associated with the projective incidence group $(G, G, \cdot)$ (compare (4.1)), then the determination of all incidence subgroups of $(G, G, \cdot)$ is equivalent to the problem of determining all near-fields lying between K and F.

In the case that $(G, \mathcal{G}, \cdot)$ is a kinematic space, each line through 1 is an incidence subgroup of dimension one and therefore the line set $\mathcal{G}(1) := \{X \in \mathcal{G}: 1 \in X\}$ is a partition of the group $(G, \cdot)$. Hence we can define a parallelism for the set $\mathcal{G}$ by calling two lines parallel if they are left cosets of the same subgroup of $\mathcal{G}(1)$. In this way, we get a parallel-structure in the sense of J. André. Moreover, in an independent generalization of affine spaces, E. Sperner introduced the concept of a weak affine space:

A triple $(P, \mathcal{G}, \|)$ is called a <u>parallel-structure</u> if $(P, \mathcal{G})$ is an incidence space (section 3) and $\|$ is an equivalence relation on $\mathcal{G}$ such that, for each point $p \in P$ and each line $A \in \mathcal{G}$, there is exactly one line $B \in \mathcal{G}$ with $p \in B$ and $B \| A$ (we denote this line by $B = \{p \| A\}$). A parallel-structure $(P, \mathcal{G}, \|)$ is a <u>weak affine space</u> if all lines of $\mathcal{G}$ have the same cardinality. Two parallel lines A, B of a parallel-structure are either the same or they have an empty intersection but the subspace generated by A and B must not be a plane as in section 3.

In [9], a quadruple $(P, \mathcal{G}, \|_\ell, \|_r)$ was called a <u>double space</u> if $(P, \mathcal{G}, \|_\ell)$ and $(P, \mathcal{G}, \|_r)$ are parallel-structures such that the following condition holds: If $A, B \in \mathcal{G}$ and $A \cap B \neq \phi$, then $\{a \|_\ell B\} \cap \{b \|_r A\} \neq \phi$ whenever $a \in A$ and $b \in B$.

From now on, let $(P, G, \|_\ell, \|_r)$ be a double space. For each $A \in G$, let $[A]_\ell := \{X \in G: X \|_\ell A\}$ and $[A]_r := \{X \in G: X \|_r A\}$.

Examples of double spaces can be obtained from kinematic spaces by taking the Clifford-parallelism: Let $(G, G, \cdot)$ be a kinematic space (section 1) and

$$\|_\ell := \{(A, B) \in G \times G;\ \exists y \in G: yA = B\}$$

(*)

$$\|_r := \{(A, B) \in G \times G;\ \exists y \in G: Ay = B\}.$$

Then $(G, G, \|_\ell, \|_r)$ is a double space; for, in a kinematic space, all lines are cosets of the lines through 1 (which are subgroups) so, if $A, B \in G$ with $A \cap B = \{c\}$ and if $a \in A$ and $b \in B$, we have $\{a \|_\ell B\} = ac^{-1}B$, $\{b \|_r A\} = Ac^{-1}b$, and $\{a \|_\ell B\} \cap \{b \|_r A\} = ac^{-1}B \cap Ac^{-1}b = \{ac^{-1}b\}$.

In [9], we proved:

(7.1) <u>All double spaces</u> $(P, G, \|_\ell, \|_r)$ <u>obtained from kinematic spaces have the following properties:</u>

$\underline{K_\ell}$ <u>Let</u> $C_1, C_2, C_3 \in [C]_r$ <u>be three distinct lines and, for each</u> $i \in \{1, 2, 3\}$, <u>let</u> $a_i, b_i \in C_i$ <u>be two points such that</u> $\overline{a_1, a_2} \|_\ell \overline{b_1, b_2}$ <u>and</u> $\overline{a_2, a_3} \|_\ell \overline{b_2, b_3}$; <u>then</u> $\overline{a_3, a_1} \|_\ell \overline{b_3, b_1}$.

K_r Let $C_1, C_2, C_3 \in [C]_\ell$ be three distinct lines and, for each $i \in \{1, 2, 3\}$, let $a_i, b_i \in C_i$ be two points such that $\overline{a_1, a_2} \parallel_r \overline{b_1, b_2}$ and $\overline{a_2, a_3} \parallel_r \overline{b_2, b_3}$; then $\overline{a_3, a_1} \parallel_r \overline{b_3, b_1}$.

A double space $(P, G, \parallel_\ell, \parallel_r)$ is called a left prism space, a right prism space, or a prism space according as K_ℓ alone holds, K_r alone holds, or both K_ℓ and K_r hold.

Moreover, the following converse of (7.1) has been established [9]:

(7.2) Let $(P, G, \parallel_\ell, \parallel_r)$ be a prism space. Then there exists a multiplication $\cdot: P \times P \to P$ such that $(P, G, \cdot)$ is a kinematic space and the parallelism $\parallel_\ell$ and $\parallel_r$ is reproduced by (*).

In a parallel-structure $(P, G, \parallel)$, one may also consider the subspaces with respect to $\parallel$: a subset $T \subset P$ is called $\parallel$-subspace if $T \in \mathcal{T}$ and, for each point $t \in T$ and each line $X \in G(T)$, the line $\{t \parallel X\}$ is contained in T. By $\mathcal{T}_\parallel$ we denote the set of all $\parallel$-subspaces of $(P, G, \parallel)$. For a double space $(P, G, \parallel_\ell, \parallel_r)$, we let $\mathcal{T}_\parallel := \mathcal{T}_{\parallel_\ell} \cap T_{\parallel_r}$. With $\mathcal{T}_\parallel$ one defines another closure operation $\hat{}$: if $M \subset P$, $\hat{M} := \cap\{T: M \subset T \in \mathcal{T}_\parallel\}$.

Since any two lines of a projective plane intersect, we cannot provide a projective plane with a parallel-structure. But this is possible for higher dimensions:

there are weak affine spaces $(P, G, \|)$ which are at the same time three-dimensional projective spaces. For instance, the kinematic space derived from the quaternions $\mathbb{H}$ over the reals $\mathbb{R}$ is both a three-dimensional projective space and a double space. The theorems (7.3) proved in [10] and (7.4) in [8] give contributions to the problem of determining all projective double spaces:

(7.3) Let $(P, G, \|_\ell, \|_r)$ be a double space such that (P, G) is a projective space and let E be a plane of the projective space (P, G). Then

(a) The closure $\hat{E}$ of E with respect to $G, \|_\ell, \|_r$ is a three-dimensional subspace of (P, G).

(b) If $\dim P = 3$, then $(P, G, \|_\ell, \|_r)$ is a projective prism space which is thus, by (7.2), derivable from a projective kinematic space.

(7.4) Every projective double space $(P, G, \|_\ell, \|_r)$ with $\|_\ell \neq \|_r$ has dimension three.

From (7.3), (7.4), and [6], we get the following major results:

(7.5) Let $(P, G, \|_\ell, \|_r)$ be a projective double space with $\|_\ell \neq \|_r$. Then there is a quaternion division-algebra (F, K) such that $(P, G, \|_\ell, \|_r)$ is derivable from the kinematic space associated with (F, K) (in the sense of section 4, Example 2(b) or Theorem (4.1)).

(7.6) Let $(P, \mathcal{G}, \|_\ell, \|_r)$ be a projective double space with $\|_\ell = \|_r$ and $\dim (P, \mathcal{G}) = 3$. Then there is a commutative field K of characteristic 2 and a pure inseparable extension field F of rank $[F: K] = 4$ such that $(P, \mathcal{G}, \|_\ell, \|_r)$ is derivable from the kinematic space associated with (F, K).

REMARK. Every pure inseparable extension field F of a commutative field K of characteristic 2 produces examples of projective double spaces with $\|_\ell = \|_r$.

In the meantime H.-J. Kroll has proved in [14a] that there are no other examples of projective double spaces with $\|_\ell = \|_r$.

Problem: Are there any (projective) incidence groups $(G, \mathcal{G}, \cdot)$ with $\underline{V_\ell}$, $\underline{V_k}$ which are not kinematic spaces (that is, where $\underline{V_r}$ is not valid)? Or in the algebraic language for the projective case: besides the kinematic division algebras, are there normal near-fields (F, K) such that $[F, K] \geq 3$ and, for each $x \in F \backslash K$, the element x^2 is contained in the set $K + Kx$?

H. Wähling has recently solved this problem for projective incidence-groups: He proved that $\underline{V_\ell}$ and $\underline{V_k}$ implies $\underline{V_r}$.

REFERENCES

1. E. Ellers and H. Karzel. Kennzeichnung elliptischer Gruppenräume. Abh. Math. Sem. Univ. Hamburg 26 (1963): 55-77.

2. ——— Endliche Inzidenzgruppen. Abh. Math. Sem. Univ. Hamburg 27 (1964): 250-264.

3. H. Karzel. Verallgemeinerte elliptische Geometrien und ihre Gruppenräume. Abh. Math. Sem. Univ. Hamburg 24 (1960): 167-188.

4. ——— Bericht über projektive Inzidenzgruppen. Jber. Deutsch. Math. Verein. 67 (1964): 58-92.

5. ——— Inzidenzgruppen I. Vorlesungsausarbeitung von I. Pieper und K. Sörensen, Hamburg (1965).

6. ——— Zweiseitige Inzidenzgruppen. Abh. Math. Sem. Univ. Hamburg 29 (1965): 118-136.

7. ——— Kinematic spaces. Istituto Nazionale di Alta Matematica, Symposia Matematica 7 (1973): 413-439.

8. H. Karzel and H.-J. Kroll. Eine inzidenzgeometrische Kennzeichnung projektiver kinematischer Räume. Arch. Math. 26 (1975): 107-112.

9. H. Karzel, H.-J. Kroll, and K. Sörensen. Invariante Gruppenpartitionen und Doppelräume. J. Reine Angew. Math. 262/263 (1973): 153-157.

10. ——— Projektive Doppelräume. Arch. Math. 25 (1974): 206-209.

11. H. Karzel and I. Pieper. Bericht über geschlitzte Inzidenzgruppen. Jber. Deutsch. Math.-Verein. 72 (1970): 70-114.

12. H. Karzel, K. Sörensen, and D. Windelberg. Einführung in die Geometrie. UTB 184, Göttingen (1973).

13. G. Kist. Punktiert-affine Inzidenzgruppen und Fastkörpererweiterungen. Abh. Math. Sem. Univ. Hamburg 44 (1975).

14. H.-J. Kroll. Zur Struktur geschlitzter Doppelräume. J. Geometry 5 (1974): 27-38.

14a. H.-J. Kroll. Bestimmung aller projektiven Doppelräume. Appearing in Abh. Math. Sem. Univ. Hamburg.

15. H. Meissner. Geschlitzte Gruppenräume. Abh. Math. Sem. Univ. Hamburg 32 (1968): 160-185.

16. H. Wähling. Darstellung zweiseitiger Inzidenzgruppen durch Divisionsalgebren. Abh. Math. Sem. Univ. Hamburg 30 (1967): 220-240.

17. —— Invariante und vertauschbare Teilfastkörper. Abh. Math. Sem. Univ. Hamburg 33 (1969): 197-202.

TOPOLOGICAL HJELMSLEV PLANES

J.W. Lorimer

Hjelmslev planes are generalizations of customary projective and affine planes where two distinct points may be joined by more than one line and two distinct lines may meet in more than one point. Two points P and Q are neighbours ($P \sim_{\mathbb{P}} Q$) when they possess more than one joining line. Two lines ℓ and m are neighbours (projectively) if they intersect in more than one point and are neighbours (affinely) if each point incident with either ℓ or m possesses a neighbour point incident with the other of ℓ or m. In these cases we write $\ell \sim_L m$.

An incidence structure $H = \langle \mathbb{P}, L, \in \rangle$ is said to be a topological Hjelmslev plane if H is a Hjelmslev plane (cf. [2], [4]) with the following properties:

(i) $\mathbb{P}$ and L are topological spaces.

(ii) $\sim_{\mathbb{P}}$ and $\sim_L$ are closed sets of $\mathbb{P} \times \mathbb{P}$ and $L \times L$ respectively.

(iii) The joining of two non-neighbouring points is a continuous function; and the meeting of two lines with a unique intersection point is a continuous function.

Moreover if the plane is affine we also assume the additional property:

(iv) L(P, g), the unique line through P parallel to g, is a continuous function of P and g.

Examples of such planes can be constructed over certain topological rings. A local ring with its unique maximal ideal equal to its set of two-sided zero divisors is an H-ring if for any two elements a and b at least one is a multiple (left and right) of the other (cf. [2]). Such a ring is a topological H-ring if it is a topological ring whose units form a multiplicative topological group. The subring D_n of the $(n \times n)$ matrices over a topological division ring F consisting of matrices of the form (dij) where dij = 0 if $i > j$ and $d_{i,i+k-1} = d_{1,k}$ is a topological H-ring (cf. [1]).

In fact from the constructions in [2], [3], and [4] we have the following result.

THEOREM 1. <u>Every desarguesian Hjelmslev plane can be coordinatized by a topological H-ring and conversely</u>.

If H is a Hjelmslev plane, then the "quotient geometry" $H/\sim \; = \; < \mathbb{P}/\sim_{\mathbb{P}}, \; L/\sim_{L}, \; \in >$ is an ordinary plane. We wish to determine under what conditions $H/\sim$ is a topological plane, when $\mathbb{P}/\sim_{\mathbb{P}}$ and $L/\sim_{L}$ are endowed with their quotient topologies. The next theorem gives necessary and sufficient conditions for H/~ to be a topological plane.

THEOREM 2. *Let* H *be a topological Hjelmslev plane. The following are equivalent:*

(a) *The quotient map* $\chi_{\mathbb{P}}: \mathbb{P} \to \mathbb{P}/\sim_{\mathbb{P}}$ *is open.*

(b) *The quotient map* $\chi_{L}: L \to L/\sim_{L}$ *is open.*

(c) *The quotient maps* $\chi_{\mathbb{P}}$ and χ_{L} *are open.*

(d) H/~ *is a topological plane.*

The following two theorems offer some sufficient conditions for H/~ to be a topological plane.

THEOREM 3. *If* H *is locally compact and connected or* H *is a translation plane, then* H/~ *is a locally compact and connected or translation topological plane.*

THEOREM 4. *If* H *is an ordered* H-*plane then* H *is a topological* H-*plane and* H/~ *is an ordered topological plane. Moreover if* H *is compact then* H *is an ordinary plane.*

REFERENCES

1. B. Artmann. Hjelmslev-Ebenen in projektive Räumen. Arch. der Math. XXL (1970): 304-307.
2. W. Klingenberg. Projektive und affine Ebenen mit Nachbarelementen. Math. Z. 60 (1954): 384-406.
3. J.W. Lorimer and N.D. Lane. Desarguesian affine Hjelmslev planes. J. Reine Angew. Math. (to appear).
4. J.W. Lorimer. Coordinate theorems in affine Hjelmslev planes. Att. Math. Pura ed Appl. (to appear).

ON FINITE AFFINE PLANES OF RANK 3

Heinz Lüneburg

We are going to study certain affine planes that admit a rank-3 collineation group. Most of these planes will turn out to be generalized André planes and, in order to establish this, we shall need a criterion that tells us that a plane is in fact a generalized André plane. To prove this criterion, we need the following lemma which was proved by Hering for finite vector spaces.

LEMMA 1. *Let* V *be a vector space over a commutative field* F *and let* A *be an abelian subgroup of* GL(V, F). *If* $V = \bigoplus_{i \in I} V_i$ *with isomorphic, irreducible* F[A]-*modules* V_i, *then* L = F[A] *is a commutative field and* V *is an* L-*vector space. Moreover* $C_{GL(V,F)}(A) = GL(V, L)$ *and* $N_{GL(V,F)}(A) \subseteq \Gamma L(V, L:F)$, *where* $\Gamma L(V, L:F)$ *denotes the group of all those bijection semilinear mappings of* V *onto itself whose companion automorphisms fix* F *elementwise. If* L *is finite, then* $N_{GL(V,F)}(A) = \Gamma L(V, L:F)$. *Furthermore* $rk_F(V) = rk_L(V)[L:F]$.

PROOF. L is of course a commutative ring. Let $0 \neq v \in V_i$, and denote by A the annihilator of v in L. Since V_i is irreducible, A is a maximal ideal of L. (Remember that L is commutative.) The irreducibility of V_i implies furthermore that $V_i A = \{0\}$. Let $z \in V_j$. Since V_j and V_i are isomorphic L-modules, there is an L-isomorphism σ from V_j onto V_i. Thus we get

$$zx = zx\sigma\sigma^{-1} = (z\sigma)x\sigma^{-1} = 0\sigma^{-1} = 0$$

for all $x \in A$. Hence $A = \{0\}$. Therefore $\{0\}$ is a maximal ideal of L. This implies, since L is a commutative ring with 1, that L is a field.

Since all F-linear mappings that centralize A centralize $L = F[A]$, we get $C_{GL(V,F)}(A) \subseteq GL(V, L)$. On the other hand we have $GL(V,L) \subseteq GL(V,F)$, because F is a subfield of L. Furthermore, $A \subseteq L^* = Z(GL(V, L))$, whence it follows that $GL(V, L) \subseteq C_{GL(V,F)}(A)$. Let $\alpha_i \in A$ and $f_i \in F$ and $\nu \in N_{GL(V,F)}(A)$. Then we get

$$\nu^{-1} \sum_{i=1}^{n} f_i \alpha_i \nu = \sum_{i=1}^{n} f_i \nu^{-1} \alpha_i \nu.$$

Hence $\nu^{-1}\lambda\nu \in L$ for all $\lambda \in L$. This implies that $\lambda \to \nu^{-1}\lambda\nu$ is an automorphism of L that fixes F elementwise. This implies that $\nu \in \Gamma L(V, L:F)$ so that $N_{GL(V,F)}(A) \subseteq \Gamma L(V, L:F)$. If L is finite, then L^* is cyclic. Therefore, A is characteristic in L^*. We conclude from this fact that $N_{GL(V,F)}(A) = \Gamma L(V, L:F)$. The proof of the fact that the F-rank of V is the product of the L-rank of V and the

F-rank of L is straightforward.

Let $Q(+, \circ)$ be a quasifield. If Q admits a second multiplication $\cdot$ such that $Q(+, \cdot)$ is a (not necessarily commutative) field and such that the mapping $x \to (x \circ a)a^{-1}$ is an automorphism of $Q(+, \cdot)$ for all $a \in Q\backslash\{0\}$, then $Q(+, \circ)$ is called a generalized André-system. A translation plane is called a generalized André-plane, if it can be coordinatized by a generalized André-system.

THEOREM 1 (Lüneburg). *Let $\mathcal{A}$ be a translation plane and let A be an abelian collineation group of $\mathcal{A}$ that fixes two points P and R on ℓ_∞ and an affine point O. If, for every point W on ℓ_∞ which is different from P and R, the stabilizer A_W of it in A induces an irreducible group of automorphisms in the group $T(W)$ of all translations with centre W, then $\mathcal{A}$ is a generalized André-plane. Moreover, if E is a point of $\mathcal{A}$ which is on neither of the lines PO and RO, and if Q is the quasifield with respect to the quadrangle P, R, O, and E, then Q is a generalized André-system.*

PROOF. It obviously suffices to prove the statement about Q. In order to do this we identify the points of $\mathcal{A}$ with the pairs $(x, y) \in Q \times Q$ such that the line OP is the set $V(0) = \{(x, 0) \mid x \in Q\}$ and the line OR is the set $V(\infty) = \{(0, x) \mid x \in Q\}$ and the point E is the point $(1, 1)$. If we denote multiplication in Q by $\circ$, then the lines through

$O = (0, 0)$ which are different from $V(\infty)$ are the point sets $V(m) = \{(x, x \circ m) \mid x \in Q\}$. Let F be the precise field of the kernel of Q. Then A is a subgroup of $GL(V, F)$, when $V = Q \oplus Q$ is to be considered as a vector space over F, the scalar multiplication being $(x, y) f = (f \circ x, f \circ y)$. We denote by A_m the stabilizer of $V(m) \cap \ell_\infty$. Our assumption implies that A_m operates irreducibly on $V(m)$ for all $m \in Q\backslash\{0\}$.

We now break up the proof into several lemmas:

LEMMA 2. _Let_ $m \in Q\backslash\{0\}$. _Then_ $F[A_m]$ _operates irreducibly on_ $V(0)$ _as well as on_ $V(\infty)$.

PROOF. Let $u \in Q\backslash\{0\}$ and $x \in Q$. Then $(u, u \circ m)$, $(x, x \circ m) \in V(m)$. Since $F[A_m]$ operates irreducibly on $V(m)$, there are $\alpha_1, \ldots, \alpha_m \in A_m$ and $f_1, \ldots, f_n \in F$ such that

$$(x, x \circ m) = \sum_{i=1}^{n} (u, u \circ m)^{\alpha_i} f_i.$$

Therefore

$$(x, 0) + (0, x \circ m) = \sum_{i=1}^{n} (u, 0)^{\alpha_i} f_i + \sum_{i=1}^{n} (0, u \circ m)^{\alpha_i} f_i.$$

Since A fixes $V(0)$ and $V(\infty)$ and since $V = V(0) \oplus V(\infty)$, we get

$$(x, 0) = \sum_{i=1}^{n} (u, 0)^{\alpha_i} f_i.$$

Thus $V(0)$ is generated as an $F[A_m]$-module by $(u, 0)$. Since this is true for all $u \in Q\backslash\{0\}$, we get that $V(0)$ is an irreducible $F[A_m]$-module. That $V(\infty)$ is an irreducible $F[A_m]$-module is proved similarly.

LEMMA 3. $V(0)$ <u>and</u> $V(\infty)$ <u>are isomorphic</u> $F[A_m]$-<u>modules</u>.

PROOF. Define the mapping μ from $V(0)$ onto $V(\infty)$ by $(x, 0)^{\mu} = (0, x \circ m)$. Then μ is an F-linear mapping from $V(0)$ onto $V(\infty)$ such that $(x, 0) + (x, 0)^{\mu} \in V(m)$. Let $\alpha \in A_m$. Then $(x, 0) + (x, 0)^{\mu} \in V(m)$ yields $(x, 0)^{\alpha} + (x, 0)^{\mu\alpha} \in V(m)$. On the other hand, $(x, 0)^{\alpha} \in V(0)$. Hence $(x, 0)^{\alpha} + (x, 0)^{\alpha\mu} \in V(m)$. Therefore

$$(x, 0)^{\mu\alpha} - (x, 0)^{\alpha\mu} \in V(m) \cap V(\infty) = \{0\}.$$

Thus $(x, 0)^{\mu\alpha} = (x, 0)^{\alpha\mu}$.

LEMMA 4. $F_m = F[A_m]$ <u>is a commutative field and</u> V <u>is a vector space of rank</u> 2 <u>over</u> F_m. <u>Furthermore</u>

$$C_{GL(V,F)}(A) = GL(V, F_m) \simeq GL(2, F_m).$$

This follows from Lemmas 1, 2, and 3.

For $\chi \in F_1$ we define $\phi(\chi)$ by $(\phi(\chi), 0) = (1, 0)^{\chi}$. Then ϕ is a bijection of F_1 onto Q. We get

$$(\phi(\chi + \lambda), 0) = (1, 0)^{\chi+\lambda} = (1, 0)^{\chi} + (1, 0)^{\lambda}$$
$$= (\phi(\chi) + \phi(\lambda), 0).$$

Hence $\phi(\chi + \lambda) = \phi(\chi) + \phi(\lambda)$ for all $\chi, \lambda \in F_1$. Define a multiplication in Q by $xy = \phi(\phi^{-1}(x)\phi^{-1}(y))$. Then we get

$$\phi(\chi\lambda) = \phi(\phi^{-1}\phi(\chi)\phi^{-1}\phi(\lambda)) = \phi(\chi)\phi(\lambda).$$

Thus ϕ is an isomorphism from F_1 onto $Q(+, \cdot)$. Furthermore

$$(x, 0)^{\chi} = (1, 0)^{\phi^{-1}(x)\phi^{-1}\phi(\chi)} = (1, 0)^{\phi^{-1}(x\phi(\chi))}$$

$$= (x\phi(\chi), 0).$$

Define ψ by $\psi(x, 0) = (0, x)$. Then ψ is an F_1-isomorphism from $V(0)$ onto $V(\infty)$. Hence

$$(0, x)^{\chi} = (\psi(x, 0))^{\chi} = \psi((x, 0)^{\chi}) = \psi(x\phi(\chi), 0)$$

$$= (0, x\phi(\chi)).$$

So we get $(x, y)^{\chi} = (x\phi(\chi), y\phi(\chi))$.

LEMMA 5. A _is a subgroup of the group_ G _of all mappings_ $(x, y) \to (xa, yb)$ _with_ $a, b \in Q\backslash\{0\}$.

PROOF. $A \subseteq C_{GL(V,F)}(A_1) = GL(V, F_1)$ by Lemma 4. Furthermore, $V(0)^A = V(0)$ and $V(\infty)^A = V(\infty)$. This yields that for every $\alpha \in A$ there are elements $\chi, \lambda \in F_1$ such that $(x, 0)^{\alpha} = (x, 0)^{\chi}$ and $(0, y)^{\alpha} = (0, y)^{\lambda}$ for all $x, y \in Q$. Hence $(x, y)^{\alpha} = (x\phi(\chi), y\phi(\lambda))$.

LEMMA 6. $F_m\backslash\{0\} \subseteq G$ _for all_ $m \in Q\backslash\{0\}$.

This follows from Lemma 5 and from $A_m \subseteq A$.

LEMMA 7. For $m \in Q\backslash\{0\}$ define $\alpha(m)$ by $x^{\alpha(m)} = (x \circ m)m^{-1}$. _Then_ α _is a mapping from_ $Q\backslash\{0\}$ _into_ $\mathrm{Aut}(Q(+, \cdot))$. _Moreover_ F_m _consists of all the mappings_ $(x, y) \to (xa, ya^{\alpha(m)})$ _where_ a _is an element of_ Q _and_ F_m _is isomorphic to_ $Q(+, \cdot)$

PROOF. Since m is different from 1, the mapping $\alpha(m)$ is a permutation of Q. Ovbiously $(1, m) \in V(m)$. Let $a \in Q$. Then $(a, a \circ m) \in V(m)$. Since $V(m)$ is an F_m-subspace of rank 1, there is a $\chi \in F_m$ such that $(a, a \circ m) = (1, m)^{\chi}$. By Lemma 6, there are elements $u, v \in Q$ such that $(x, y)^{\chi} = (xu, yv)$ for all $x, y \in Q$. Thus $(a, a \circ m) = (u, mv)$ and hence $a = u$ and $v = a^{\alpha(m)}$, proving that $((x, y) \to (xa, ya^{\alpha(m)})) \in F_m$. Conversely if $\chi \in F_m$, then $(1, m)^{\chi} = (a, a \circ m)$ for some $a \in Q$ and we get $\chi = ((x, y) \to (xa, ya^{\alpha(m)}))$. Now

$$\begin{aligned}(x + y)^{\alpha(m)} &= ((x + y) \circ m)m^{-1} \\ &= (x \circ m + y \circ m)m^{-1} \\ &= (x \circ m)m^{-1} + (y \circ m)m^{-1} \\ &= x^{\alpha(m)} + y^{\alpha(m)}.\end{aligned}$$

Finally, if $(x, y)^{\chi} = (xa, ya^{\alpha(m)})$, $(x, y)^{\lambda} = (xb, yb^{\alpha(m)})$ and $(x, y)^{\chi\lambda} = (xc, yc^{\alpha(m)})$, then $(xab, ya^{\alpha(m)}b^{\alpha(m)}) = (x, y)^{\chi\lambda} = (xc, yc^{\alpha(m)})$, which yields $c = ab$ and $a^{\alpha(m)}b^{\alpha(m)} = (ab)^{\alpha(m)}$.

It follows from Lemma 7 that $Q(+, \circ)$ is a generalized André-plane. Thus Theorem 1 is proved.

The converse of Theorem 1 is not true, since it is easily seen that a desarguesian plane over a non-commutative field does not admit such an abelian collineation group. I do not know whether a translation plane over a generalized André-system $Q(+, \circ)$ such that $Q(+, \cdot)$ is a commutative field always admits such an abelian collineation group. However, every finite generalized André-plane admits such a group. It is easily seen that the group described by Ostrom [17, Theorem 1] has all the derived properties. If one checks this, one has to keep in mind that there is no generalized André-plane of order 2^6 such that the kernel of the plane is GF(2) (Foulser [5, Lemma 3.4]).

An affine plane is called a rank-3-plane if it admits a collineation group that acts transitively on its point set such that the stabilizer of a point has exactly three orbits. Kallaker [10] and Liebler [13] proved that a finite affine plane A admitting a rank-3-collineation group G is a translation plane and that G contains the translation group of A. It is easily seen that G has at most two orbits on the line at infinity. If G has two orbits on the line at infinity, then G_ℓ is 2-transitive on the set of points on ℓ for all lines ℓ of A.

The following theorem is due to Kallaker, Ostrom, and Lüneburg [11], [12], [14], [15], and [16].

THEOREM 2. *Let* A *be a finite affine plane of order* n. *If* A *admits a rank-3-collineation group* G *such that* G *has an orbit of length 2 on* ℓ_∞, *then* A *is a generalized André-plane or* $n \in \{5^2, 7^2, 11^2, 23^2, 29^2, 59^2\}$.

The numbers listed in the theorem are real exceptions since all the irregular near fields yield planes that satisfy the hypotheses of the theorem, but none of these planes is a generalized André-plane. We shall give more information about the exceptions later on.

As we have pointed out, A is a translation plane and G contains the translation group T of A. Let 0 be an affine point of A. Then $G = TG_0$. Put $H = G_0$. Since T induces the identity on ℓ_∞, the group G and H have the same orbits on ℓ_∞. Let $\{P, Q\}$ be the orbit of length two on the line at infinity of the group H. Our concern is to show that $H_{P,Q}$ contains an abelian subgroup that satisfies the hypothesis of Theorem 1.

H has three point orbits. One of them is $\{0\}$. The second one consists of the points on OP and OQ other than O. Its length is $2(n - 1)$. The third orbit consisting of the remaining points has length $n^2 - 2(n - 1) - 1 = (n - 1)^2$. Thus the order of H is divisible by 2 and by $(n - 1)^2$. Furthermore, $|H: H_{P,Q}| = 2$.

Since A is a translation plane, n is a power of a prime p. Let $n = p^r$. A prime π is called a p-primitive prime divisor of $p^r - 1$ if π divides $p^r - 1$ but if π does not divide $p^i - 1$ for all i with $1 \leq i < r$. We consider now the case where there exists a p-primitive prime divisor π of $p^r - 1$. Let π^s be the highest power of π dividing $p^r - 1$. Then π^{2s} divides $|H|$. Let π^{2s+t} be the highest power of π dividing $|H|$. Obviously, π cannot be 2 except in the case $r = 1$. But in this case, A is desarguesian and hence a generalized André-plane. Thus we can assume $\pi \neq 2$. Since $|H: H_{P,Q}| = 2$, we get that π^{2s+t}

divides $|H_{P,Q}|$. Put $K = H_{P,Q}$; denote by $K(P, OQ)$ the group of all (P, OQ)-homologies contained in K and define $K(Q, OP)$ similarly. Let Σ be a Sylow-π-subgroup of K. Since $K(P, OQ)$ is normal in K, the group $\Sigma \cap K(P, OQ)$ is a Sylow-π-subgroup of $K(P, OQ)$. Furthermore $|\Sigma \cap K(P, OQ)| \leq \pi^s$, since $(K(P, OQ))$ divides $p^r - 1$. Let X be an affine point of OQ that lies in an orbit of length π^s of Σ. Such an orbit exists, since K is transitive on the set of affine points of OQ which are different from O. Thus $|\Sigma : \Sigma_X| = \pi^s$. Let $\sigma \in \Sigma_X$ such that σ^π induces the identity on OQ. Let C be the centralizer of σ in $T(Q)$. Then $C \neq \{1\}$ because of $O \neq X = X^\sigma$. Put $|C| = p^\alpha$. If $\alpha \neq r$, then $\langle\alpha\rangle$ splits $T(Q) - C$ into orbits of length π. This implies that π divides $|T(Q) - C| = p^r - p^\alpha = p^\alpha(p^{r-\alpha} - 1)$. Therefore π divides $p^{r-\alpha} - 1$, a contradiction, since $1 \leq \alpha < r$. Hence Σ_X induces the identity on OQ. Therefore $\Sigma_X \subseteq \Sigma \cap K(P, OQ)$. Now

$$\pi^{s+t} = |\Sigma_X| \leq |\Sigma \cap K(P, OQ)| \leq \pi^s$$

so $\Sigma_X = K(P, OQ) \cap \Sigma = \Sigma(P, OQ)$ and $|\Sigma(P, OQ)| = \pi^s$. Likewise, we get $|\Sigma(Q, OP)| = \pi^s$. Since $\Sigma(P, OQ) \cap \Sigma(Q, OP) = \{1\}$, we have furthermore that $\Sigma = \Sigma(P, OQ)\Sigma(Q, OP)$.

It is easily verified that $T(P)\Sigma(P, OQ)$ is a Frobenius group and $\Sigma(P, OQ)$ is a Frobenius complement of it. Since $\Sigma(P, OQ)$ is a π-group and π is odd, $\Sigma(P, OQ)$ is cyclic (see, for example, Huppert [9, (8.15a)]). The group $\Sigma(Q, OP)$ is cyclic for the same reason. Thus Σ is abelian

Assume that $\Sigma(P, OQ)$ is normal in $K(P, OQ)$. Then $\Sigma(P, OQ)$ is the unique Sylow-π-subgroup of $K(P, OQ)$. Since $\{P, Q\}$ is an orbit of H, the groups $K(P, OQ)$ and $K(Q, OP)$ are conjugate in H. Therefore $\Sigma(Q, OP)$ is the only Sylow-π-subgroup of $K(Q, OP)$. This implies that Σ is normal in H. Therefore, all the orbits of Σ in $\ell_\infty \setminus \{P, Q\}$ have the same length, which is obviously π^s. This implies that $|\Sigma_W| = \pi^s$ for all points W on ℓ_∞ other than P and Q. Let $1 \neq \sigma \in \Sigma_W$ be an element of order π and let C be the centralizer of σ in T(W). As above, $C = \{1\}$ or $C = T(W)$. If $C = T(W)$, then σ fixes all points on OW, since it fixes O. Hence σ is a perpectivity with axis OW and centres P and Q. This yields $\sigma = 1$, a contradiction. Hence $C = \{1\}$. It follows therefore that the group of automorphisms induced by Σ_W on T(W) is isomorphic to Σ_W. Again using the fact that π is a p-primitive prime divisor of $p^r - 1$, we discover that Σ_W operates irreducibly on T(W). Therefore, A is a generalized André-plane by Theorem 1.

Next we need the following:

LEMMA 8. *Let* V *be an elementary abelian p-group of order* p^r *and assume that* π *is a p-primitive prime divisor of* $p^r - 1$. *Assume furthermore that* Σ *is a non-trivial* π-*subgroup of* GL(r, p) *and that* N *is an abelian subgroup of* GL(r, p) *that operates regularly on* $V\setminus\{1\}$. *If* Σ *normalizes* N, *then* ΣN *is cyclic*.

PROOF. Given a non-trivial π-subgroup of GL(r, p), then it operates irreducibly on V. Since it is contained in the centralizer of its centre, it must be cyclic by Schur's lemma. Thus Σ is cyclic and operates irreducibly on V. Since N is a Frobenius complement by assumption, N is also cyclic.

Assume that π divides $|N|$. If N_0 is the Sylow π-subgroup of N, then $N_0\Sigma$ is a π-subgroup of GL(r, p) that normalizes N. Since $N_0\Sigma$ is a cyclic π-group by our previous remark, we get $\Sigma \subseteq N_0$ or $N_0 \subseteq \Sigma$. If $\Sigma \subseteq N_0$ there is nothing to prove. So let $N_0 \subseteq \Sigma$. Thus $\Sigma \cap N \neq \{1\}$. Furthermore $\Sigma N \subseteq C_{\Sigma N}(\Sigma \cap N)$. By Schur's lemma, $C_{\Sigma N}(\Sigma \cap N)$ is cyclic. Hence we are done in this case.

Assume that π does not divide $|N|$ and put $|N| = n$. Let Σ_0 be the subgroup of order π of Σ. If Σ_0 centralizes N, then $\Sigma N \subseteq C_{\Sigma N}(\Sigma_0)$. Using Schur's lemma again, we obtain that ΣN is cyclic. Thus we only have to show that Σ_0 centralizes N. We now make an induction on n. If n = 1, then Σ_0 centralizes N. Let $n > 1$. If $n = ab$ with $a \neq 1 \neq b$ and $(a, b) = 1$, then $N = N_aN_b$ where N_a is a subgroup of order a and N_b is a subgroup of order b. Both of them are centralized by Σ_0 by the induction hypothesis and hence N is. Thus we may assume $n = a^t$ where a is a prime. Let $N = \langle\nu\rangle$ and $\Sigma_0 = \langle\sigma\rangle$. Assume furthermore that Σ_0 does not centralize N. Then $\nu^\sigma = \nu^i$ where $(i - 1, a^t) = 1$ and $i^\pi \equiv 1 \bmod a^t$ (Passman [18, Prop. 12.11]). By the induction hypothesis, σ centralizes ν^a. Thus $\nu^a = \nu^{a\sigma} = \nu^{ai}$, whence $i \equiv 1 \bmod a^{t-1}$. Since a does not divide i - 1, we get t = 1. This shows that $\Sigma_0 N$ is a Frobenius group with Frobenius kernel N and Frobenius complement Σ_0. Since a Frobenius group of order $\pi \cdot a$ cannot operate regularly on $V \backslash \{1\}$ (Huppert [9, (8.15.b)]), there is an element $\xi \neq 1$ in $\Sigma_0 N$ and a $\nu \in V\backslash\{1\}$ such that $v = \nu^{\xi}$. Now $o(\xi) = a$ or π since $\Sigma_0 N$ is a Frobenius group. But π-elements operate regularly on $V\backslash\{1\}$. Thus $\xi \in N$, a contradiction.

COROLLARY. *Let* V *be an elementary abelian p-group of order* p^r *and assume that* π *is a* p-*primitive prime divisor of* $p^r - 1$. *If* N *is a subgroup of* GL(r, p) *all of whose Sylow subgroups are cyclic and if* N *operates regularly on* $V\backslash\{1\}$, *then* N *contains exactly one Sylow* π-*subgroup*.

PROOF. This is true if π does not divide $|N|$. Assume that π divides $|N|$. By Passman [18, Prop. 12.11], there is a cyclic normal subgroup M of N and a cyclic subgroup C such that N = MC and $(|M|, |C|) = 1$. If π divides $|M|$, then π does not divide $|N/M|$. Thus we are done in this case. Assume that π divides C. Then C contains exactly one Sylow π-subgroup Σ of N. Furthermore, $M\Sigma$ contains the commutative subgroup of N, since N/M is abelian. Thus $M\Sigma$ is normal in N. Since $M\Sigma$ is cyclic by Lemma 8, the corollary follows.

Now, $K(P, OQ)$ is a Frobenius complement of the Frobenius group $T(P)K(P, OQ)$. This yields:

(A) If $K(P, OQ)$ is solvable, then $K(P, OQ)$ contains a normal subgroup N all of whose Sylow subgroups are cyclic such that $K(P, OQ)/N$ is isomorphic to a subgroup of S_u.

(B) If $K(P, OQ)$ is non-solvable, then $K(P, OQ)$ contains a subgroup L of index 1 or 2 and L is the direct product of a group $M \simeq SL(2,5)$ and a group N such that $|M|$ and $|N|$ are relatively prime and such that the Sylow subgroups of N are all cyclic.

A proof of these facts can be found in Passman [18, Sect. 18].

Assume $\pi \neq 3$ and assume that $K(P, OQ)$ is solvable if $\pi = 5$. In this case we infer from (A), (B), and Lemma 9 that $\Sigma(P, OQ)$ is normal in $K(P, OQ)$ and hence that A is a generalized André-plane. Thus we may assume that there is no p-primitive prime divisor of $p^r - 1$ which is different from 3 and 5, and that $K(P, OQ)$ is non-solvable in the case that 5 is a p-primitive prime divisor of $p^r - 1$.

If 3 is a p-primitive prime divisor of $p^r - 1$, then $r = 2$, since $p^2 \equiv 1 \bmod 3$. If 5 is a p-primitive prime divisor of $p^r - 1$, then $r = 2$ or $r = 4$, since $p^2 \equiv 1 \bmod 5$ or $p^4 \equiv 1 \bmod 5$. If $r = 4$, then $p^2 + 1$ is divisible by 5. Furthermore, $p \neq 2, 3$, since $|K(P, OQ)|$ divides $p^4 - 1$ and since $|SL(2, 5)| = 120$. Thus $p^2 - 1 \equiv 0 \bmod 12$ and, hence, $p^2+1=2.5^t$. This yields that 5^t divides $|\Sigma(P, OQ)|$ which is 5 by (5). Therefore $p^2 + 1 = 10$ and $p = 3$. Thus 120 divides $3^4 - 1 = 80$, a contradiction. So we get $r = 2$ in either case.

LEMMA 9. *Let* A *be a translation plane of order* p^2 *where* p *is a prime, and let* G *be its collineation group. Assume that there are two points* P, Q *on* ℓ_∞ *and an affine point* O *such that* $G(P, OQ)$ *contains a subgroup* S *isomorphic to* $SL(2, 5)$. *If* G_O *operates transitively on the set of affine points on* OP *other than* O, *then* $p = 11, 19, 29, 59$.

PROOF. $G(P, OQ)$ is a Frobenius complement of the Frobenius group $T(P)G(P, OQ)$. Hence S is the only subgroup of $G(P, OQ)$ isomorphic to $SL(2, 5)$ by (B). Therefore, S is normal in G_O. Since G_O is transitive on $OP\cdot\{O\}$, we get that G_O operates transitively on $T(P)\setminus\{1\}$. Then G_O maps onto a subgroup $\bar{G}_O$ of $PGL(2, p)$. Furthermore, S is mapped onto a normal subgroup $\bar{S}$ of $\bar{G}_O$ which is isomorphic to A_5. This implies that $\bar{S} = \bar{G}_O$ according to Dickson [4, §259]. Thus $p + 1$ divides 60. This yields $p = 2, 3, 5, 11, 19, 29, 59$. Since 120 divides $p^2 - 1$, we obtain $p = 11, 19, 29, 59$.

Thus, if $\pi = 3$ or 5 and if $K(P, OQ)$ is non-solvable, then $p^r = 11^2, 19^2, 29^2, 59^2$ are the only possible exceptions. All these numbers except $p = 19$ are listed in Theorem 2. The case $p = 19$ must be handled separately;

more will be said about this case later on.

We may assume from now on that $\pi = 3$ is the only p-primitive prime divisor of $p^r - 1$ and that $K(P, OQ)$ is solvable. Furthermore, we may assume that $\Sigma(P, OQ)$ is not normal in $K(P, OQ)$. The next lemma will take care of this situation.

LEMMA 10. *Let* A *be a translation plane of order* p^2 *where* p *is a prime and let* G *be its collineation group. Assume that there are two points* P, Q *on* ℓ_∞ *and an affine point* O *such that* $G(P, OQ)$ *is solvable and that* $G(P, OQ)$ *contains more than one Sylow* 3-*subgroup. If* G_O *operates transitively on the set of affine points on* OP *other than* O, *then* $p = 5, 7, 11, 23$.

PROOF. G_O maps onto a subgroup $\bar{G}_O$ of $PGL(2, p)$ which operates transitively on the projective line over $GF(p)$. Furthermore, $G(P, OQ)$ maps onto a normal subgroup $\bar{M}$ of $\bar{G}_O$. It follows by our assumption on the Sylow 3-groups of $G(P, OQ)$ and by Dickson's list of the subgroups of $PGL(2, p)$ that $\bar{M}$ contains a subgroup $\bar{A}$ isomorphic to A_n such that $|\bar{M} : \bar{A}| \leq 2$. Thus $\bar{A}$ is characteristic in $\bar{M}$ and hence normal in $\bar{G}_O$. It follows from Dickson [4, §257] that $[\bar{G}_O : \bar{A}] \leq 2$ and hence that $|\bar{G}_O|$ divides 24. Therefore $p + 1$ is a divisor of 24 which yields $p = 2, 3, 5, 7, 11, 23$. Since 12 divides $p^2 - 1$, we get $p = 5, 7, 11, 23$.

We get from Lemma 10 that $p^r = 5^2, 11^2, 23^2$ are the

only possible exceptions if 3 is the only p-primitive prime divisor of $p^r - 1$.

We now assume that there is no p-primitive prime divisor of $p^r - 1$. A famous theorem of Zsigmondy thus tells us that $p^r = 2^6$ or that $r = 2$ and $p + 1$ is a power of 2 (Zsigmondy [19]; Zsigmondy's theorem is also in Artin [1]).

Assume that $p^r = 2^6$. We want to show that $K(P, OQ) \neq \{1\}$. Thus, suppose that $K(P, OQ) = \{1\}$. Since H is transitive on the 63^2 points of the lines OP and OQ and since $|H: K| = 2$, we get that K is also transitive on the set of these points. Hence 63^2 divides $|K|$. Let K* be the group induced by K on OQ. Since $K(P, OQ) = \{1\}$, we have $K^* \simeq K$. Thus $63^2 = 3^4 \cdot 7^2$ divides $|K^*|$. As a result, we get $K^* = GL(6, 2)$ by Hering [8, Lemma 5.9]. This implies that K contains an involution σ that centralizes exactly 2^5 elements in T(Q). We infer from this fact that σ fixes exactly 2^5 points on OQ, because σ fixes O. On the other hand, σ being an involution implies that σ is a central collineation or a Baer involution. In either case we get the contradiction that σ does not fix exactly 2^5 points on a line. Thus $K(P, OQ) = \{1\}$.

The fact that $K(Q, OP) \simeq K(IQ, OP)K(P, OQ)(K(P, OQ))$ implies that K* contains a normal subgroup Z which is isomorphic to K(Q, OP). Let $|Z|$ be a prime. Then $|Z| = 3$ or 7 and therefore Z operates reducibly on T(Q). Since K* is transitive on $T(Q)\backslash\{1\}$, we obtain that the Z-irreducible subgroups of $T(Q)$ form a geometric partition. Baer [2, sect. 2] then implies that $K^* \subseteq \Gamma L(3, 4)$ if $|Z| = 3$ and that $K^* \subseteq \Gamma L(2, 8)$ if $|Z| = 7$. Now $|\Gamma L(3, 4)| = 2^7 \cdot 3^4 \cdot 5 \cdot 7$ and $|\Gamma L(2, 8)| = 2^3 \cdot 3^3 \cdot 7^2$. Thus we get a contradiction: $3 = |Z| = |K(Q, OP)| = |K(P, OQ)| = 7$ or $7 = |Z| = |K(Q, OP)| = |K(P, OQ)| = 3$ respectively. This contradiction shows that $|K(Q, OP)| \in \{9, 21, 63\}$.

If $|K(Q, OP)| = 63$, then $\mathcal{A}$ is desarguesian or a near-field plane. In either case, $\mathcal{A}$ is a generalized André-plane. Next let us consider the case $|K(Q, OP)| = 9$. Then $|K(P, OQ)K(Q, OP)| = 81$ and $A = K(P, OQ)K(Q, OP)$ is normal in H. Thus all the orbits of A on $\ell_\infty \backslash \{P, Q\}$ have the same length, a say. This length a divides 63 and $|A|$ and is divisible by 9. Hence $a = 9$. Therefore $|A_W| = 9$, if W I ℓ_∞ and $W \neq P, Q$. The group A_W operates faithfully on OW and hence the group of automorphisms induced by A_W in $T(W)$ operates faithfully on $T(W)$. Furthermore, $K(P, OQ)$ and $K(Q, OP)$ are cyclic according to Huppert [9, (8.15a)]. This implies that A_W, as a diagonal of $K(P, OQ)K(Q, OP)$, is also cyclic. We infer from all this that A_W operates irreducibly on $T(W)$. Thus $\mathcal{A}$ is a generalized André-plane by Theorem 1.

Let $|K(Q, OP)| = 21$. Since $21 = 3 \cdot 7$, we get that $K(Q, OP)$ is cyclic (Huppert [9, (8.15b)]). Therefore $A = K(P, OQ)K(Q, OP)$ is abelian. Again, all the orbits of A on $\ell_\infty \backslash \{P, Q\}$ have the same length a, and we get a = 21 or 63. If a = 21, the above argument gives that A is a generalized André-plane. So let us assume that a = 63. We shall show that this cannot happen.

Let W I g_∞ and W $\neq$ P, Q; then $|A_W| = 7$. Let S be a Sylow 3-subgroup of A. It follows from $|A_W| = 7$ that S operates regularly on $\ell_\infty \backslash \{P, Q\}$. Since $K(P, OQ) \simeq K(Q, OP)$, the group K* has a cyclic normal subgroup of order 21 that operates free of fixed points on $T(Q)\backslash\{1\}$ and hence irreducibly on T(Q). As a result, $K^* \subseteq \Gamma L(1, 64)$ by Lemma 1. We know that $|\Gamma L(1, 64)| = 2 \cdot 3^3 \cdot 7$. Furthermore 63^2 divides $|K| = |G(P, OQ)|\ |K^*| = 21|K^*|$. Thus $|K| = 63^2$ or $2 \cdot 63^2$. Hence K contains a normal subgroup B of order 63^2. Since K is transitive on the 63^2 points of the lines OP and OQ, we deduce that B is also transitive and, hence, sharply transitive on these points. Therefore, B_W operates sharply transitively on $OW\backslash\{O, W\}$, whence it follows that $T(W)B_W$ is a Frobenius group with Frobenius complement B_W. Let Σ be a Sylow 3-subgroup of B_W. Then Σ is cyclic of order 9. Furthermore, $S \cap \Sigma = \{1\}$, since S operates regularly on $\ell_\infty \backslash \{P, Q\}$. Thus $|S\Sigma| = 9^2$ so that $S\Sigma$ is a Sylow 3-subgroup of K. Let S_1 be a Sylow 3-subgroup of K(P, OQ) and S_2 be a Sylow 3-subgroup of K(Q, OP). Thus $S = S_1S_2$. Moreover, S_i is normalized by Σ and hence centralized, since $|S_i| = 3$ and $|\Sigma| = 9$. Thus $S \subseteq Z(S\Sigma)$ and $\Sigma S/S$ is cyclic. Hence $S\Sigma$ is abelian, a contradiction, because the Sylow 3-

subgroups of $\Gamma L(1, 64)$ are not abelian.

We are left with the case that the order of $\mathcal{A}$ is p^2 and that $p + 1 = 2^s$. To handle this case we need two lemmas.

LEMMA 11. *Let $\mathcal{A}$ be a translation plane of order p^2, where p is a prime such that $p \equiv 3 \bmod 4$. Let P, Q be two points on ℓ_∞ and let O be an affine point of $\mathcal{A}$. If G is a collineation group of $\mathcal{A}$ fixing P, Q, and O and if G is transitive on the set of affine points of the lines OP and OQ, then the order of a Sylow 2-subgroup of $G(P, OQ)$ is 2^s or 2^{s+1} where 2^s is the highest power of 2 dividing $p + 1$. If the order of a Sylow 2-subgroup of $G(P, OQ)$ and of a Sylow 2-subgroup of $G(Q, OP)$ is 2^s, then the Sylow 2-subgroups of either group are cyclic.*

PROOF. Let Σ be a Sylow 2-subgroup of G. By our assumption, $(p^2 - 1)^2$ divides $|G|$. Hence $|\Sigma| = 2^{2s+2+b}$ where b is a non-negative integer. Σ normalizes $T(P)$, since P is fixed by G. Let Σ^* be the centralizer of $T(P)$ in Σ. Then Σ/Σ^* is isomorphic to a subgroup of $GL(2, p)$. It follows from $O^{\Sigma^*} = O$ that $\Sigma^* \subseteq G(Q, OP)$. On the other hand, $G(Q, OP)$ centralizes $T(P)$. Thus $\Sigma^* = \Sigma \cap G(Q, OP)$, whence it follows that Σ^* is a Sylow 2-subgroup of $G(Q, OP)$, because $G(Q, OP)$ is normal in G.

Let $|\Sigma^*| = 2^t$. Then $t \leq s + 1$, since 2^{s+1} is the highest power of 2 dividing $p^2 - 1$. On the other hand,

2^{s+2} is the highest power of 2 dividing $p(p - 1)^2(p + 1)$ $= |GL(2, p)|$. Thus

$$2^{2s+2+b-t} = |\Sigma/\Sigma^*| \leq 2^{s+2}.$$

This yields $s \leq s + b \leq t \leq s + 1$, which proves the first assertion.

Let $t = s$. Thus $b = 0$ and $|\Sigma| = 2^{2s+2}$. Furthermore, let $\Sigma^{**} = G(P, OQ) \cap \Sigma$. Then $|\Sigma^{**}| = 2^s$ or 2^{s+1}, since the assumptions are symmetric in P and Q. Assume $|\Sigma^{**}| = 2^s$. The group Σ^{**} is the centralizer in T(Q) in Σ. Thus Σ/Σ^{**} is isomorphic to a subgroup of GL(2, p). We infer from $|\Sigma/|**| = 2^{s+2}$ that Σ/Σ^{**} is isomorphic to a Sylow 2-subgroup of GL(2, p). It follows from Carter and Fong [3, p. 142] that Σ/Σ^{**} is a semidihedral group, because $p \equiv 3$ mod 4. The group $\Sigma^*\Sigma^{**}/\Sigma^{**}$ is a normal subgroup of index 4 in Σ/Σ^{**}. Using the generators and relations for a semidihedral group (see, for example, Gorenstein [6, p. 191]), one obtains that all elements of order 2 which do not lie in the maximal cyclic subgroup are conjugate and that the same is true for the corresponding elements of order 4. From this it follows that there is only one normal subgroup of index 4 in a semidihedral group, i.e. the cyclic subgroup of index 2 is the maximal cyclic subgroup. Hence $\Sigma^*\Sigma^{**}/\Sigma^{**}$ is cyclic. Now $\Sigma^* \simeq \Sigma^*\Sigma^{**}/\Sigma^{**}$, proving the lemma.

LEMMA 12. *Let* $\mathcal{A}$ *be a translation plane of order* p^2 *where* p *is a prime such that* $p \equiv 3$ mod 4. *Let* P, Q *be two points* on ℓ_∞ and let O be an affine point of A. If G is a collineation group of A fixing P, Q, and O and if G splits the set of points of the lines PO and OQ into two orbits of length $\frac{1}{2}(p^2 - 1)^2$, then the order of a Sylow 2-subgroup of G(P, OQ) is 2^{s-1}, 2^s, or 2^{s+1}, where 2^s is the highest power of 2 dividing p + 1. If it is 2^{s-1} and if the order of a Sylow 2-subgroup of G(Q, OP) is also 2^{s-1}, then the Sylow 2-subgroups of G(P, OQ) are cyclic.

The proof of Lemma 12 is similar to that of Lemma 11.

Now we come back to the proof of Theorem 2. We have that the order of A is p^2 when $p + 1 = 2^s \geq 4$. It follows from (B) and Lemma 9 that K(P, OQ) is solvable, and Lemmas 11 and 12 imply that a Sylow 2-subgroup of K(P, OQ) has order 2^{s-1}, 2^s, or 2^{s+1} and that the Sylow 2-subgroups of K(P, OQ) are cyclic, if their order is 2^{s-1}, since K(P, OQ) and K(Q, OP) are conjugate. If s = 2, then p = 3. By M. Hall [7, Appendix II], there are only two translation planes of order 9, the desarguesian one and the plane over the near-field plane of order 9. Both planes are generalized André-planes. If x = 3, then p = 7, one of the exceptions. Thus we may assume that $s \geq 4$, which implies $s \geq 5$.

By the remark that K(P, OQ) is solvable and by (A), there is a normal subgroup N of K(P, OQ) all of whose Sylow subgroups are cyclic such that K(P, OQ)/N is isomorphic to a subgroup of S_4. We show that the Sylow subgroups of odd order of N are all in the centre of K(P, OQ). Let a be the biggest prime that divides $|N|$. If $a = 2$, there is nothing to prove. Assume that $a > 2$. All Sylow subgroups of N being cyclic, a theorem of Burnside (see, for example, Gorenstein [6, Theorem 7.4.3]) implies that the Sylow a-subgroup A of N is normal in N and therefore normal in K(P, OQ). Since $|K(P, OQ)|$ divides $p^2 - 1 = 2^s(p - 1)$, we get that $|A|$ divides $p - 1$. Therefore A operates reducibly on T(P). Let Σ be a Sylow 2-subgroup of K(P, OQ). Then $|\Sigma| > 2^{s-1} > 16$. Furthermore, the stabilizer of a subgroup of order p of T(P) in Σ has order 2, because $p \equiv 3 \bmod 4$. Hence A, being normalized by Σ, fixes at least 8 subgroups of order p of T(P). This is more than enough to ensure that A is in the centre of GL(2, p) and hence of K(P, OQ). We conclude from the above-mentioned theorem of Burnside that A has a normal a-complement B in N. As a Hall subgroup of N, the group B is characteristic in N and hence normal in K(P, OQ). Induction now yields the desired result.

Put $\Sigma_0 = \Sigma \cap N$. Then Σ_0 is a Sylow 2-subgroup of N. Since all Sylow subgroups of odd order of N are in the centre of K(P, OQ), we get that Σ_0 is normal in N and hence in K(P, OQ). Furthermore, N is cyclic.

Assume that Σ is not normal in K(P, OQ). Then 3 divides $|K(P, OQ)/N|$. Let M/N be the largest normal 2-subgroup of K(P, OQ)/N and let Σ_1 be a Sylow 2-subgroup of M.

Let C be the complement of Σ_0 in N. Then $M = \Sigma_1 C$ and $\Sigma_1 \supseteq \Sigma_0$.

Since C is in the centre of K(P, OQ), we get that Σ_1 is normal in M and therefore in K(P, OQ). Assume $M \neq N$. Then $|\Sigma_1/\Sigma_0| = 4$, since K(P, OQ)/N is isomorphic to a subgroup of S_4 and since 3 divides |K(P, OQ)/N|. Furthermore, K(P, OQ) induces a group of automorphisms on Σ_1 whose order is divisible by 3. Now, Σ_1 is a generalized quaternion group. But the only generalized quaternion group that admits an automorphism of order 3 is the quaternion group of order 8. Hence $|\Sigma_1| = 8$. Since Σ_1 is not cyclic, Σ is not cyclic. Therefore, $|\Sigma| = 2^s$ or $|\Sigma| = 2^{s+1}$. Furthermore, $|\Sigma: \Sigma_1| = 2$, because Σ is not normal in K(P, OQ). This gives the contradiction $16 = |\Sigma| \geq 2^s = 32$. Thus M = N and $\Sigma_0 = \Sigma_1$. From this it follows $|\Sigma: \Sigma_0| = 2$ and hence $|\Sigma_0| \geq 2^{s-1}$. Obviously, Σ_0 is the intersection of all the Sylow 2-subgroups of K(P, OQ). Therefore Σ_0 is a characteristic subgroup of K(P, OQ). Since Σ_0 is cyclic and since $|\Sigma_0| \geq 2^{s-1}$, this implies that there is a characteristic cyclic subgroup Σ^* of K(P, OQ) such that $|\Sigma^*| = 2^{s-1}$.

Let Σ be normal in K(P, OQ). Then Σ contains also a characteristic subgroup Σ^* of order 2^{s-1}. This is true by Lemma 12, if $|\Sigma| = 2^{s-1}$. If $|\Sigma| \geq 2^s$, then Σ is either cyclic or a generalized quaternion group of order ≥ 16. In either case, Σ contains such a subgroup Σ^*. Since Σ is characteristic in K(P, OQ), the group Σ^* is also characteristic.

Let Σ be a characteristic cyclic subgroup of order 2^{s-1} of K(P, OQ). Since K(Q, OP) is conjugate to K(P, OQ) there is also a characteristic cyclic subgroup Σ^{**} of order 2^{s-1} in K(Q, OP). Obviously, Σ^* and Σ^{**} are unique. Hence $A = \Sigma^*\Sigma^{**}$ is a normal subgroup of H. (Remember that $H = G_O$ where G is the rank-3-group under consideration.) Since Σ^* and Σ^{**} centralize each other and since $\Sigma^* \cap \Sigma^{**} = \{1\}$, the group A is abelian of order 2^{2s-2}. A being normal in H implies that all the orbits of A in $\ell_\infty \backslash \{P, Q\}$ have the same length. Thus all these orbits have length at most 2^{s+1}. Let $W \mathrm{I} \ell_\infty$ and $W \neq P, Q$. It follows then that

$$2^{2s-2} = |A| = |W^A|\,|A_W| \le 2^{s+1}|A_W|.$$

Therefore $|A_W| \ge 2^{s-3} \ge 4$. The centralizer of T(W) in A_W consists of homologies with axis OW and centres P and Q, whence it follows that it is equal to {1}. This means that A_W is isomorphic to a group of automorphisms of T(W). The group A being the direct product of two cyclic groups contains only three involutions. Thus A_W contains only one involution. Therefore A_W is cyclic. Thus A_W induces a cyclic group Γ of automorphisms on T(W). Since $|\Gamma| = |A_W| \ge 4$ and since $p \equiv 3 \bmod 4$, the group Γ operates irreducibly. This finally implies, using Theorem 1, that A is a generalized André-plane.

We shall finish this paper by a few comments on the exceptional orders that occur in the theorem and about the case 19^2 which has not been ruled out so far. If one has got an exception to the theorem of order different from 7^2, then it can be shown that this exception belongs to one of the following two classes.

THE CLASS R * p. Let A be an affine plane of order p^2 where p is a prime. A is said to be of type R * p if it satisfies the following conditions:

(i) If G is the collineation group of A and if ℓ is any line of A, then G_ℓ operates 2-transitively on ℓ.

(ii) There are two points P, Q I ℓ_∞ and a point O $\not I$ ℓ_∞ such that G(P, OQ) contains a subgroup S_1 and G(Q, OP) contains a subgroup S_2 where $S_1 \simeq S_2 \simeq SL(2, 5)$. Furthermore, the set of orbits of S_1 on ℓ_∞ is the same as the set of orbits of S_2 on ℓ_∞.

THE CLASS F * p. Let A be an affine plane of order p^2 where p is a prime. A is said to be of type F * p if it satisfies the following conditions:

(i) same as for R * p.

(ii) There are two points P, Q I ℓ_∞ and a point O $\not I$ ℓ_∞ such that G(P, OQ) contains a normal subgroup S_1 and G(Q, OP) contains a normal subgroup S_2 such that $S_1 \simeq S_2 \simeq SL(2, 3)$. Furthermore, the set of orbits of S_1 on ℓ_∞ is the same as the set of orbits of S_2 on ℓ_∞.

It turns out that p = 11, 19, 29, or 59, if A is of type R * p. Furthermore, there is exactly one plane of type R $\star$ 11 and exactly one of R $\star$ 59. There are exactly 3 planes of type R $\star$ 19 and exactly 9 of type R $\star$ 29. The number of exceptions to the order 2 in each class is 1, 1, 0, 1 respectively.

If A is of type F $\star$ p, then p = 5, 7, 11, 23. The number of planes in each class is 1, 2, 4, 1 and the number of exceptions is 1, 2, 1, 1 respectively. There may be more exceptional planes of order 7^2. For details see Lüneburg [15].

REFERENCES

1. E. Artin. The order of the linear groups. Comm. Pure and Appl. Math. 8 (1955): 355-365.
2. R. Baer. Partitionen abelscher Gruppen. Arch. Math. 14 (1963): 73-83.
3. R. Carter and P. Fong. The Sylow-2-subgroups of the finite classical groups. J. Algebra 1 (1964): 139-151.
4. L.E. Dickson. Linear groups. New York: Dover Publications (1958).
5. D. Foulser. A generalization of André's systems. Math. Z. 100 (1967): 380-395.
6. D. Gorenstein. Finite Groups. New York, Evanston and London: Harper and Row (1968).
7. M. Hall. Projective planes. Trans. Am. Math. Soc. 54 (1943): 229-277.
8. C. Hering. Transitive linear groups and linear groups which contain irreducible groups of prime order. Geometriae Dedicata 2 (1974): 425-460.
9. B. Huppert. Endliche Gruppen I. Berlin-Heidelberg-New York: Springer Verlag (1967).
10. M.J. Kallaker. On finite affine planes of rank 3. J. Algebra 13 (1969): 544-553.
11. —— A note on Z-planes. J. Algebra 28 (1974): 311-318.
12. M.J. Kallaker and T.G. Ostrom. Fixed point free linear groups, rank three planes, and Bol quasifields. J. Algebra 18 (1970): 159-178.
13. R.A. Liebler. Finite affine planes of rank 3 are translation planes. Math. Z. 116 (1970): 89-93.

14. H. Lüneburg. Über eine Klasse von endlichen affinen Ebenen des Ranges 3. To be published in "Atti del Colloquio Internazionale sulle Teorie Combinatorie." The conference in question took place in Rome, Sept. 3-15, 1973.
15. —— Über einige ruschwerdige Translations-ebenen. Geometriae Dedicata (to appear).
16. H. Lüneburg and T.G. Ostrom. Affine Ebenen des Ranges 3 mit einer Bahn der Länge 2 auf der uneigentlichen Geraden. Forthcoming.
17. T.G. Ostrom. A characterization of generalized André planes. Math. Z. 110 (1969): 1-9.
18. D.S. Passman. Permutation groups. New York: W.A. Benjamin Publ. Comp. (1968).
19. K. Zsigmondy. Zur Theorie der Potenzreste. Monatshefte der Math. und Phys. 3 (1892): 265-284.

EVERY GROUP IS THE COLLINEATION GROUP OF SOME PROJECTIVE PLANE[1]

E. Mendelsohn

We shall use in this note graph-theoretical results and a process adopted from category theory to translate these results into the following results of projective geometry.

We shall prove the following four theorems.

THEOREM 1. *If* G *is a group, then there exists a projective plane* $\mathcal{P} = (P, L, I)$ *such that the collineation group of* $\mathcal{P}$, $A(\mathcal{P}) \simeq G$, *and if* α *is a cardinal*, $\alpha \geq \aleph_0\, |\mathcal{P}| = \alpha|G|$.

THEOREM 2. *If* G_i, $\{i \in I\}$ *is a well-ordered sequence of groups, then there exist projective planes* $\mathcal{P}_i$ *such that* $\mathcal{P}_i$ *is a subplane of* $\mathcal{P}_j$ *whenever* $i < j$ *and* $A(\mathcal{P}_i) \simeq G_i$.

THEOREM 3. *Let groups* G_1 *and* G_2 *be given. Then there exist projective planes* $\mathcal{P}_1 = (P_1, L_1, I_1)$ *and* $\mathcal{P}_2 = (P_2, L_2, I_2)$ *and a pair of onto maps* $\phi: P_1 \to P_2$ *and* $\psi: L_1 \to L_2$ *such that* $pI_1\ell \Rightarrow \phi(p)I_\alpha\psi(\ell)$ (*an onto homomorphism* [6]) *with*

1. This is a summary of "Every group is the collineation group of some projective plane" and "Pathological projective planes," both of which appeared in Journal of Geometry.

$A(P_1) \simeq G_1$ and $A(P_2) \simeq G_2$.

THEOREM 4. Let G be a group. Then there exists a projective plane p such that for every normal subgroup N of G (including 1 and G) there is a line ℓ, such that if one takes ℓ the line at infinity, the collineation group of the affine plane, $p_{\ell=\infty}$, is N.

RESULTS NEEDED FROM GRAPH THEORY

DEFINITION. The category of I-multicoloured graphs $R(2_i)_{i \in I}$, I a set, is defined as follows:

Objects: A set x together with a family of subsets $R_i \subset X \times X$, $i \in I$.

Morphisms: $f:(X, R_i) \to (Y, S_i)$ $(i \in I)$ is a morphism if $f: X \to Y$ is a function and for all $i \in I$ $(x, x') \in R_i \Rightarrow (f(x), f(x')) \in R_i$.

We shall refer to $R(2_i)_{i \in \{1\}}$ as R(2) and call it the category of graphs. The elements of R_i are called i-(coloured)-arrows. We shall say "f preserves R_i" if $f: x \to y$ is a morphism from (X, R_i) to (Y, S_i) in R(2). A loop of R_i is a point x such that $(x, x) \in R_i$, an isolated point of R_i is a point such that $y(x, y) \notin R_i$ and $(y, x) \notin R_i$; a two-cycle in R_i is a pair (x, y) such that (x, y) and $(y, x) \in R$.

$\mathrm{Aut}(X, R_i)_{i \in I} = \{f \mid f$ and $f^{-1} \in \mathrm{Hom}\,(X, R_i)_{i \in I}, (X, R_i)_{i \in I})\}$; $(X, R_i)_{i \in I}$ is rigid if $\mathrm{Hom}\,(X, R_i)_{i \in I}(X, R_i)_{i \in}$

$\simeq 1_x$. $\mathrm{Stab}_{(x,y)}(X, R_i)_{i\in I} = \{f \mid f\ \mathrm{Aut}(X, R_i)_{i\in I} f(x) = x\ f(y) = y\}$.

THEOREM 5 (Cayley-Frucht). Let G be a group and $(X, R_g)_{g\in G}$ be a graph with $|G|$ colours defined by $X = |G|$ and $(g_1, g_2) \in R_h \iff g_1 h^{-1} = g_2$. Then $\mathrm{Aut}(X, R_g)_{g\in G} = G$ [1].

LEMMA 1. Let G be a group. Then there exists a multi-coloured graph $(X, R_i)_{i\in I}$ such that

(i) $\mathrm{Aut}(X, R_i)_{i\in I} \simeq G$.

(ii) For $N < G$ (including $N = 1$ and $N = G$) $(x, y) \in R_{i(N)}$ such that $\mathrm{Stab}_{(x,y)}((X, R_i)_{i\in I}) \simeq N$.

CONSTRUCTION. Let $I = \{a, b, c\} \dot{\cup} \{G \times \mathcal{N}\}$ where $\mathcal{N} = \{N \mid N \Delta G$ ($N = 1$ and $N = G$ included)$\}$. Let

$X = \{\frac{G}{N} \mid N < G\} \dot{\cup} \{Hg \mid Hg \in \frac{G}{H}, H < G\}$,

$R_a = \{$a rigid graph on $\{\frac{G}{N} \mid N < G\}\}$,

$R_b = \{(\frac{G}{N}, Ng) \mid Ng \in \frac{G}{N}\}$,

$R_c = \{(Ng, Kh) \mid Ng \subseteq Kh \text{ and } N \subseteq K\}$,

$R_{(N,g)} = \{(Nh, Nh') \mid Nh = Nh'g\}$.

LEMMA 2. If G is a multicoloured graph, then there exists a graph G_1 with the same automorphism group, and the effect of stabilizing an arrow of G can be achieved by stabilizing an arrow of G_1 [4].

THEOREM 6. If G is a group, then there exists a graph (X, R) such that $A(X, R) \simeq G$, and if $\alpha \geq \aleph_0$ is a cardinal $|X| = \alpha|G|$ [3], [4].

THEOREM 7. *If G_i, $i \in I$ is a well-ordered sequence of groups, then there exist graphs* (X_i, R_i) *with* $X_i \subset X_j$, $R_i \subset R_j$, $i < j$, *and* $A(X_i, R_i) \simeq G_i$ [8].

THEOREM 8. *If G_1, G_2 are groups then there exist graphs* (X_1, R_1) *and* (X_2, R_2), *and an arrow-onto map* $\phi: (X_1, R_1) \to (X_2, R_2)$ *such that* $A(X_1, R_1) \simeq G_1$ *and* $A(X_2, R_2) \simeq G_2$ [5].

Furthermore by a technique found in [4] and [8] one can assume that all graphs which appear in the above theorems have no loops, isolated points, or two-cycles, and that each point is related to at least two others. These assumptions are for technical reasons which will reveal themselves in the particular construction and proofs we shall use.

RESULTS NEEDED FROM PROJECTIVE GEOMETRY

We shall need the following well-known results from projective geometry:

THEOREM 9. *The collineation group of* $p_{\ell=\infty}$ *is isomorphic to the subgroup of the collineation group stabilizing* ℓ [7].

THEOREM 10. *Given a partial plane* p *there exists a free completion of this plane to a projective plane* [2].

THEOREM 11. *If a partial plane is confined* (*i.e. contains at least three points on each line and at least three lines* through every point) then the free completion has the same collineation group as the partial plane and furthermore the isomorphism is given by restriction [2].

The configuration $\mathcal{D}$

The configuration $\mathcal{D}$ is given by the following table where the columns are lines and the entries points:

A	C	E	A	B	D	D	C	E	A	B
B	N	O	F	K	N	O	K	M	K	N
C	L	L	G	L	K	M	G	G	O	O
D	H	F	H	M	F	H				
E										

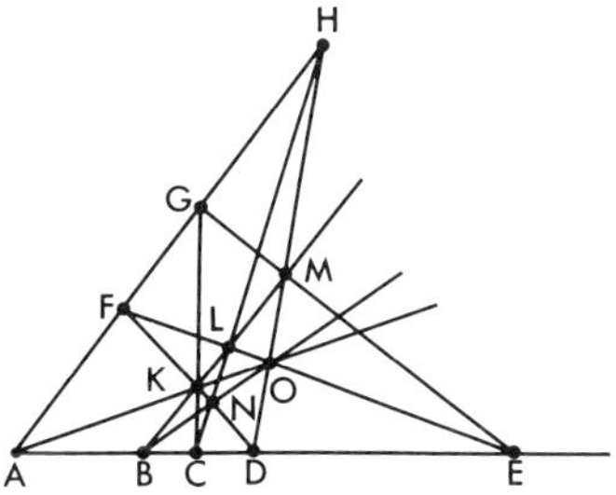

FIGURE 1

LEMMA 3. $\mathcal{D}$ has no collineations other than the identity and is confined.

The pasting of $\mathcal{D}$ on a graph

The process we shall use is analogous to the $\check{s}$ip process [8]. Let G = (X, R) be a graph without loops, isolated points, or two-cycles and with each point connected to at least two others. We define the configuration (G) in the following way:

The points of $\mathcal{D}(G)$: $P(G) = \left\{[\{P_D - \{K, O\}\} \times R] \cup X\right\}$.

The lines of $\mathcal{D}(G)$: $L(G) = L_D \times R$.

The incidence of $\mathcal{D}(G)$: I_G by:

(i) $(p, r)\ I_G\ (\ell, s) \iff r = s$ and $p\ I_D\ \ell$;

(ii) $x\ I_G\ (\ell, s) \iff s = (x, y)$, $K\ I_D\ \ell$;

(iii) $y\ I_G\ (\ell, s) \iff s = (y, z)$, $O\ I_D\ \ell$.

Intuitively, we place a copy of $\mathcal{D}$ in place of every arrow of R pasting the points K and/or O together whenever they correspond. If we represent the configuration $\mathcal{D}$ as in figure 2 and paste it on the graph of figure 3 we get the configuration of figure 4.

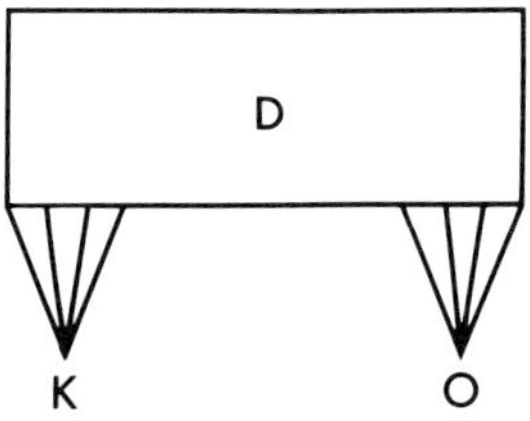

FIGURE 2

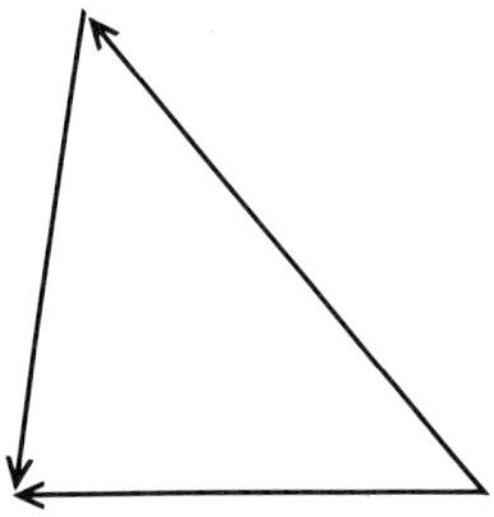

FIGURE 3

THEOREM 12. If G = (X, R) is a graph such that G has no loops or isolated points, and every point is connected to at least two others, then:

1. $\mathcal{D}(G)$ is a confined configuration.

2. There is an isomorphism between the collineation

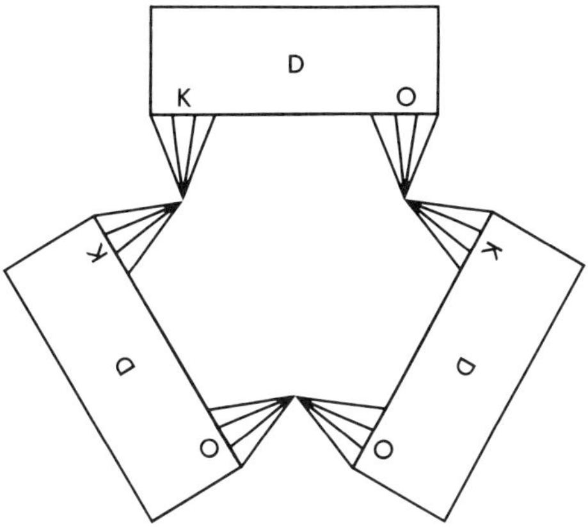

FIGURE 4

group of $\mathcal{D}$(G) and the automorphism group of the graph G.

3. There is a one-one correspondence between onto maps from G to G* (where G* also has the properties given in the hypothesis) and onto homomorphism from $\mathcal{D}$(G) to $\mathcal{D}$(G*).

4. Stabilizing an arrow of G corresponds to stabilizing a line of $\mathcal{D}$(G).

Theorems 1-4 now follow from the analogous theorems for graphs and considering the plane F($\mathcal{D}$(G)).

REFERENCES

1. R. Frucht. Herstellung von Graphen mit vorgegeben abstrakter Gruppe. Composito Math. 6 (1938): 239-250.
2. M. Hall. Projective planes. Trans. Amer. Math. Soc. 54 (1943): 229-277.
3. Z. Hedrlin, A. Pultr, and P. Vopenka. A rigid relation exists on any set. Comment. Math. V. Carolinae 6 (1965): 149-155.
4. Z. Hedrlin and J. Lambek. How comprehensive is the category of semigroups? J. Algebra 11, No. 2 (1969): 195-212.

5. Z. Hedrlin. On endomorphisms of graphs and their homomorphic images. In: Proof Techniques in Graph Theory, edited by F. Harary. Academic Press (1969).
6. D.R. Hughes. On homomorphism of projective planes. Proc. Symp. Appl. Math. 10 (1960): 45-52.
7. R. Lingenberg. Grundlagen der Geometry 1. Bibliographische Institute, Zürich (1969).
8. E. Mendelsohn. On a technique for representing semigroups as endomorphism semigroups of graphs with given properties. Semigroup Forum 4 (1972): 283-294.

RECENT ADVANCES IN FINITE TRANSLATION PLANES

T.G. Ostrom

I. GROUPS GENERATED BY GIVEN TYPES OF COLLINEATIONS

The work on rank-three affine planes, reported on by Lüneburg at this same conference, is certainly a phase of the "recent advances" but I shall make no further mention of it here. Most of what I have to say was reported on at a Conference on Projective Planes held at Washington State University. Copies of the Proceedings may be obtained from the Washington State University Press. My report in those Proceedings contains what I hope is a complete listing of known finite translation planes and some of the properties of the collineation groups. Since then I believe that Rao has some more examples of flag transitive planes. Prohaska and Walker have an example, not yet published, of case (e) in the theorem I shall soon be stating. I believe that the list is complete with these additions. Some of what I shall talk about is in my "Finite Translation Planes."

If π is a translation plane of dimension r over its kernel K, then the (affine) points of π can be identified with the elements of a 2r-dimensional vector space over the field K; the lines through the zero vector O are r-dimensional vector subspaces and the scalar transformations are the homologies with centre O and axis ℓ_∞. The system of lines through the origin is called a spread.

Although some of what we have to say does not require finiteness, we shall assume that everything is finite. "Point" always means "affine point" unless we say "point at infinity." The letter q will denote the order of K: K = GF(q). The stabilizer of O in the collineation group is a group of semilinear transformations - a subgroup of $\Gamma L(2r, q)$. We call it the translation complement and its intersection with GL(2r, q) the linear translation complement. The characteristic of K will be denoted by p.

The first result, I am told, is implied by some unpublished work of John Thompson. In the following form it is due in part to Hering and in part to myself.

THEOREM. Suppose that the translation complement contains (affine) elations. Let S be the subgroup of the translation complement which is generated by all of the elations. Then one of the following holds:

(a) S is an elementary abelian p-group and (if S is non-trivial) only one line through O is the axis of a non-trivial elation.

(b) $S \simeq SL(2, p^s)$ *for some* s. *The net whose lines through the origin are the axes of elations can be embedded in a desarguesian affine plane of order* q^r.

(c) $p = 2$ *and* $S \simeq Sz(2^s)$ *for some* s.

(d) S *contains a normal subgroup* N *of odd order and index two in* S.

(e) $p = 3$ *and* $S \simeq SL(2, 5)$. *We have a desarguesian net as in case* (b).

Details of the proof are given in [7], [13], and [18]. The first part is concerned with the group generated by two elations with different axes. It depends upon the fact that the group generated by two matrices $\begin{pmatrix}1 & 0\\1 & 1\end{pmatrix}$ and $\begin{pmatrix}1 & a\\0 & 1\end{pmatrix}$ over a field of characteristic $p \neq 3$ is $SL(2, p^s)$, where $GF(p^s)$ is the smallest subfield containing a. A representation can be used so that two elations with different axis can be represented by $\rho_1 = \begin{pmatrix}I & 0\\I & I\end{pmatrix}$ and $\rho_2 = \begin{pmatrix}I & A\\0 & I\end{pmatrix}$, where I is an r by r identity matrix and A is an r by r submatrix with an eigenvalue a.

Let K' be an extension of the field K which contains the eigenvalue a of A and let $W \oplus W$ be a vector space of dimension 2r over K'. If we think of the elements of W as ordered r-tuples of elements of K' then the vector space $V \oplus V$ obtained by restricting the r-tuples to elements of K can be identified with the one in which the plane π is represented. Furthermore $\langle \rho_1, \rho_2 \rangle$ acts as a reducible group of linear transformations on $W \oplus W$. There is an invariant subspace of dimension two on which it turns out to act faithfully. The restriction to this subspace is conjugate to the group generated by $\begin{pmatrix} 1 & 0 \\ 1 & 1 \end{pmatrix}$ and $\begin{pmatrix} 1 & a \\ 0 & 1 \end{pmatrix}$. From this, one concludes that $\langle \rho_1, \rho_2 \rangle \simeq SL(2, p^s)$. For $p = 3$, there is the additional possibility that $\langle \rho_1, \rho_2 \rangle \simeq SL(2, 5)$.

For a given elation σ in the translation complement, Hering denotes by $E(\sigma)$ the group of elations with the same axis as σ and $\Sigma = \{E(\sigma)\}$. If, for some $E, F \in \Sigma$, $\langle E, F \rangle = SL(2, p^s)$, then $\langle E, F \rangle$ can be represented by a set of matrices $\begin{pmatrix} A & B \\ C & D \end{pmatrix}$ such that the ring generated by all of the submatrices is a field K, and $AD - BC = I$. Thus the r by r submatrices are identified with elements of a field so that $\begin{pmatrix} A & B \\ C & D \end{pmatrix}$ can be interpreted as a two by two matrix.

Using the knowledge about the group generated by a pair of elements and letting $\langle E, F \rangle = \langle \sigma, \tau \rangle$ we get a field of matrices $K(\sigma, \tau)$. For given σ, $\bigcup_{\tau \in F} K(\sigma, \tau)$ is a field. From this, Hering deduces that $\langle E, F \rangle \simeq SL(2, p^s)$ for each distinct E, F in Σ and finally arrives at conclusion (b) if $p > 3$.

Some modifications are required for characteristic 2 or 3. Note the remark we have appended to the listing for [18].

A remark or two about this theorem. The second part of conclusion (b) suggests that one might be able to construct examples by modifying a desarguesian plane. I have been able to carry this out [15].

Let G be the (linear) translation complement (all elations in the translation complement will be in the linear translation complement) and let G_1 be the subgroup generated by all the elations. As a matter of pure group theory, if $G \triangleright G_1$, then G modulo $C(G_1)$ (the centralizer in G of G_1) is isomorphic to the group of automorphisms of G_1 induced by taking conjugates. In the present case, $C(G_1)$ is very restricted since it must leave each elation axis invariant. In fact it is precisely the group fixing all elation axes.

We would like to find other theorems of this type for planes in which the translation complement contains no elations.

For large dimension it may be helpful to try to characterize what I call minimal non f.p.f. groups.

DEFINITION. A group of linear transformations is fixed-point-free if no non-trivial element of the group fixes any non-zero vector.

DEFINITION. A normal subgroup G_1 of a group G of linear transformations is said to be minimal non-f.p.f. with respect to G if:

(a) G_1 is not f.p.f. (some non-trivial element of G_1 fixes some non-zero vector).

(b) If H is any normal subgroup of G properly contained in G_1, then H is f.p.f. The minimal property ensures that if σ is any non-f.p.f. element of G_1, then G_1 is generated by all of the conjugates of σ with respect to G. See [17].

Foulser [3] has shown that for $p > 3$ the arguments used on groups generated by elations work as well for groups generated by Baer p-elements. A Baer p-element is an element of order p that fixes pointwise a subplane whose order is the square root of the order of the full plane.

Thus we have the following problem: determine the groups generated by Baer p-elements for $p = 2$ or 3 subject to the restriction that the group contains no elations.

Most of the really interesting known finite translation planes are of dimension two over the kernel, i.e., the vector space has dimension four. I think it would be a big advance if we could list all of the possibilities for dimension two. This can now be done, in principle, for planes of even order since Wagner, in a paper not yet published, has determined all subgroups of GL(2, q) for q even. It remains, of course, to be determined how these groups can act on translation planes of dimension two. The task will probably be easier if we only look at minima non-f.p.f. groups.

When both the dimension and order are odd, we have a good knowledge of the abstract groups but a shortage of examples of planes and of the representations of these groups that can act as collineation groups.

There may be some point here in a thorough investigation of the small dimensions, say 3 and 5, so that the dimension of the vector space is 6 or 10 respectively.

Let G be the linear translation complement and let $\bar{G}$ be the induced group on the line at infinity. Thus $G \subset GL(2r, q)$ and $\bar{G}$ is the image of G in the homomorphism onto $PGL(2r, q)$. If both r and p are odd, a result of Hering [8] says that the Sylow 2-groups in $\bar{G}$ are cyclic or dihedral. By use of the Gorenstein-Walter theorem, I have been able to show that the non-solvable minimal non-f.p.f. groups with respect to G are isomorphic to SL(2, u) for some odd u except possibly when A_7 is a factor group. In my original version I also included PSL(2, u) as a possibility, but Lüneburg has pointed out that this case can be eliminated because PSL(2, u) contains too many involutions. Nothing here tells us that u must be a power of the characteristic. If u is not a power of p, a result of Hering and Harris [4] implies that $u \leq 2d + 1$ where d is the dimension of the vector space. For planes of dimension 3, d is 6 and the minimal u is 13. Hering's planes of order 27 do admit SL(2, 13) [5]. I know of no other example of a translation plane of odd order and dimension which admits a non-solvable collineation group. Hering has another result which shows that the stabilizer of a line is always solvable in this case.

In another direction, we might ask: "What about groups generated by homologies with affine axis?" More specifically, suppose that P and Q are points on ℓ_∞ and there is a non-trivial (P, OQ) homology of order u. It can happen that P and Q are in a single orbit of length two or that both are fixed by the translation complement. It also might happen that the subgroup generated by conjugates of the given homology contains elations. These are all of the possibilities in a desarguesian plane unless u = 2, 3, or 5 in which case we can get a dihedral group, SL(2, 3) or SL(2, 5). I have been able to show that all of this is still true for translation planes of odd dimension [16]. Let me remind you that two homologies with the same centre and different axes (or the same axis and different centres) generate a group which contains elations.

The proof depends on looking at groups generated by a pair of homologies and which are, in a certain sense, minimal but such that the orbit of P contains points besides P and Q. Neglecting a few complications which arise if P and Q are in the same orbit we get Frobenius groups with complement of order u if the group is solvable. This turns out to be possible only for u = 2 or 3.

For characteristic two, if the group is non-solvable

it must contain involutions. At odd dimension the involutions are elations; if there are no elations the group is solvable by Feit-Theompson.

For both the solvable and non-solvable case, we do what amounts to reducing the problem to the desarguesian case: we represent the two generating matrices in the form $\begin{pmatrix} A_1 & B_1 \\ C_1 & D_1 \end{pmatrix}$ and $\begin{pmatrix} A_2 & B_2 \\ C_2 & D_2 \end{pmatrix}$ and show that the submatrices A_1, B_1, ..., C_2, D_2 generate a ring which is a field. The Gorenstein-Walter theorem gives us possibilities for the permutation group induced on the orbit of the centre P. With a little case-by-case argument, we can show that we always can choose a representation so that the submatrices commute and do generate a field.

Our conjecture for translation planes of dimension two is: If P is in an orbit of length greater than two (where there is a (P, OQ) homology of order u) and if the group generated by homologies of order u contains neither elations nor Baer p-elements, then u = 2, 3, or 5.

II. REDUCIBILITY CONSIDERATIONS

In the general theory of linear groups, the groups acting on vector spaces of relatively small dimension are less complicated than when the dimension is higher. When the group is reducible one is enabled to restrict the consideration to groups of smaller dimension. We shall also call attention to certain other notions that amount to a kind of reducibility. We shall illustrate this approach by looking at the case where the dimension is already small. For larger dimensions, there are more cases to consider but the same methods can be applied in principle. In general, if σ is an element of the linear translation complement, the set of points fixed by σ is a vector subspace, which we shall denote by $V(\sigma)$. Furthermore either $V(\sigma)$ is a subspace of a line through the origin or is the set of points of an affine subplane. In the latter case, the dimension of $V(\sigma)$ must be even, say 2t, and the intersection of $V(\sigma)$ with each line through the origin is a vector subspace and the dimension is either zero or t.

Now, specifically, suppose that G is the linear translation complement or a subgroup of the linear translation complement of a finite translation plane of dimension two over K, i.e., the vector space has dimension 4; the subspaces of dimension two are lines through the origin or are Baer subplanes. We can make a sort of classification in terms of reducibility as follows:

1. G has an invariant subspace on which it does not act faithfully. The subgroup fixing this subspace pointwise is normal in G.

(a) G has precisely one invariant line ℓ through O and G does not act faithfully on ℓ. Note that ℓ is the axis of a non-trivial group of elations. All of the semifield planes are of this kind.

(b) G has precisely two invariant lines through 0 and G does not act faithfully on one or both. Here G contains affine homologies. There are many examples of this kind.

(c) G either has no invariant lines through the origin or at least acts faithfully on the lines it does fix but G has a unique invariant Baer subplane and does not act faithfully on this subplane. In this case we have a p-group fixing the Baer subplane pointwise. Some planes of this type are derived from those of type 1(a).

(d) Faithful on all invariant lines, a unique pair of invariant Baer subplanes. Not faithful on at least one of these. Analogous to case (b).

(e) Faithful on all invariant two-spaces but there are invariant 1-spaces on which the action is not faithful. I know of no examples of planes of this kind.

2. G is reducible but acts faithfully on all of its invariant subspaces. There must be at least one invariant line or at least one invariant Baer subplane. The collineation group is isomorphic to some subgroup of GL(2, q). The groups generated by elations (with more than one elation axis) are of this type (for odd characteristic) but the full translation complement is irreducible in the known cases.

DEFINITION. A group G of non-singular linear transformations operating irreducibly on a vector space V is _imprimitive_ if $V = V_1 \oplus \ldots \oplus V_k$ and for each $i = 1, \ldots, k$ and each σ in G there is a j such that $V_i\sigma = V_j$. If no such decomposition is possible G is said to be _primitive_.

Thus an imprimitive group is irreducible but is very much like a reducible one and, in fact, contains a normal reducible group of relatively small index. For instance: if P and Q are points on ℓ_∞ such that every element of G fixes or interchanges P and Q and if G is irreducible then G is imprimitive.

3. G is irreducible but imprimitive and does not act faithfully on its subspaces of imprimitivity. All of the near-field planes and most of the generalized André planes are of this type. That is, we have (P, OQ) homologies and P, Q are in an orbit of length 2. There will be an analogous situation in some of the planes derived from these in which the subspaces of imprimitivity are Baer subplanes.

4. As in 3 but the action is faithful on the subspaces of imprimitivity.

It may happen that G is irreducible and even primitive but is reducible in some other sense. Suppose that our basic vector space V be regarded as the space of ordered 4-tuples over the kernel K and that K' is an extension of K. Let W be the vector space of 4-tuples over K'. If G is represented by matrices with elements in K, G acts on W. It can happen that G is reducible or imprimitive on W. It turns out that if G is primitive on V it must act faithfully on its irreducible constituents in W. It also turns out that no 3-dimensional subspace of W can be an irreducible constituent of G. Hence G must be isomorphic to a subgroup of $GL(2, q^i)$ for some i; in fact G could be isomorphic to a 1-dimensional linear group - but see below.

DEFINITION. If G is irreducible over every extension K' of K, then G is said to be absolutely irreducible.

There is another way of effectively reducing the dimension. Let H be an abelian group acting irreducibly on a vector space V of dimension d over K. (Here we do not need d = 4.) Then H is cyclic and the elements of V can be identified with the elements of $GF(q^d)$ in such a way that H is represented by multiplication maps $x \to xa$ in the larger field.

This generalizes as follows. Let G be a primitive subgroup of GL(n, q) and let A be a maximal abelian normal subgroup of G. Suppose that $(|A|, q) = 1$ and A is f.p.f. Then A is cyclic and the ring of endomorphisms generated by A is a field K. If $|K| = q^t$ then $t \mid n$, V is a vector space of dimension n/t over a field isomorphic to K, and G is isomorphic to a subgroup of $\Gamma L(n/t, q^t)$. If G as a subgroup of GL(n, q) is primitive, then A corresponds to a group of scalar matrices in $\Gamma L(n/t, q^t)$.

Although it does not directly apply here, we might mention the following as stated in Dixon [1], Corollary 4.2A. If G is an irreducible primitive linear group over an algebraically closed field, then every normal abelian subgroup consists of scalars and is in the centre.

This kind of reduction of dimension is, in certain cases, closely related to certain methods by which planes are constructed. This may, in some sense, be considered to be a version of Cliffords theorem (see Dixon [1]). Hering uses a slightly different version in [9]. A proof of the present version is given in [19].

In [11] and [14], I attempted to take a broad look at constructions. Whether their originators looked at them that way or not, most of the successful constructions of translation planes can be looked upon as examples of one or the other of the two folllowing general methods or combinations of the two.

1. To construct a plane admitting a translation complement acting in a certain way with a small number of orbits, get an explicit representation of what the group must be like. By usually ad hoc or trial-and-error methods choose certain initial lines through the origin; the others are defined as their images.

2. Modify some known plane. Usually the original

plane can be taken to be desarguesian and the modification involves replaceable nets. The replacements very often amount to the following: In a desarguesian affine plane of order q^r coordinatized by $GF(q^r) = \hat{K}$, think of $\hat{K}$ as a vector space of dimension r over $K = GF(q)$ so that the plane is a vector space of dimension 2r over GF(q). Let W be a vector subspace of dimension r such that for each $a \neq 0$ in $\hat{K}$ the image of W under the mapping $(x, y) \rightarrow (xa, ya)$ is either equal to W or intersects W only in the zero vector. We now replace the lines through 0 which intersect W with W and its images under this cyclic group of order $q^r - 1$. This group continues to act as a collineation group of the new plane and is, in fact, normal. This procedure can be carried out simultaneously in several different parts of the plane. Now let us return to our earlier considerations. Planes constructed in this way will admit the kind of reduction we have just been discussing.

Suppose that we have a translation plane of dimension r over GF(q) and that the translation complement has a normal abelian subgroup H such that $V = V_1 \oplus V_2$ where V_1 and V_2 are r-dimensional subspaces left invariant by H and irreducible under H. Suppose also that the linear translation complement is irreducible and primitive. Then V has an image different from V_1 or V_2. In fact V is a union of H-irreducible subspaces of dimension r which can be thought of as lines through the origin of a desarguesian affine plane and H can be identified with an irreducible group of scalar maps $(x, y) \to (xa, ya)$. The given plane can be constructed from the desarguesian plane by a modification of the procedure we have been describing.

REMARK. These ideas are very close to the ones which led to my characterization of generalized André planes [12] and the improvement by Lüneburg. (See Lüneburg's lecture at this same conference.) An equivalent characterization was made independently by Rao, Rodabaugh, Wilke, and Zemmer [22]. The known planes of odd order which contain elations appear to have the property that the translation complement is irreducible and primitive but that a minimal non-f.p.f. group is reducible - i.e. there is an invariant subplane. The full linear translation complement is subject, however, to the kind of reduction mentioned above. This is almost equivalent to saying that these planes were constructed from desarguesian planes.

If a group is absolutely irreducible and if its order is relatively prime to the characteristic, it is isomorphic to a group of linear transformations on a vector space of the same dimension over the field of complex numbers. The primitive complex groups are known [2] for small dimensions. I have examined those of dimension 4 [19]; some of them cannot act as collineation groups of translation planes of dimension two.

All of this survey does not quite amount to a complete explicit listing of possibilities. For instance, we have not tried to list all planes of dimension two with a pair of points P, Q on ℓ_∞ in a single orbit of length two such that there is a non-trivial (P, OQ) group of homologies.

Nevertheless, all possibilities but one are included in some sense. Let me try to explain this last case. For a plane of dimension two an element of order p in the translation complement must be of one of three types: (a) elation, (b) Baer p-element, (c) the pointwise fixed subspace is 1-dimensional. The missing case is the one in which we have an absolutely irreducible group in which the p-elements are of this third type. We do have examples due to Hering [6] and the guess is that others which might exist all have essentially the same nature.

A similar analysis can be carried out, say, for planes of dimension three over the kernel - the vector space has dimension 6. The number of ways a p-element can act, and the possible geometric nature, are somewhat greater.

The number of examples of known planes is much smaller. Here is the place where we can use more examples of translation planes. We do, however, have much more knowledge as to the possible abstract groups which can appear.

III. ORDERS OF ELEMENTS

In this section we are interested in the possible orders of elements of the translation complement and how they act. Again, we shall illustrate what can actually be a more general approach by reference to the case of planes of dimension two.

Now $|GL(4, q)| = q^6(q + 1)^2(q - 1)^4(q^2 + 1)(q^2 + q + 1)$ so that if σ is an element of prime order, then $|\sigma|$ divides $p(q - 1)(q + 1)(q^2 + 1)(q^2 + q + 1)$. Now $|\sigma|$ must divide $q^4 - 1$ if it is fixed point free. If σ is a homology, $|\sigma|$ divides $q^2 - 1$; if for some line ℓ through the origin, the dimension of $V(\sigma)$ ℓ is one, then $|\sigma|$ divides $q^2 - q = q(q - 1)$.

Thus the order of every non-f.p.f. element (if prime) must divide $q^2 - 1$. Conversely, suppose that $|\sigma|$ does divide $q^2 - 1$. Then σ fixes at least two lines through the origin. If $|\sigma|$ divides $q + 1$ but $|\sigma| \neq 2$ and σ leaves some 1-space pointwise invariant. The number of 1-spaces in a two space is $q + 1$ and if σ fixes two 1-spaces pointwise in the same two-space it must fix this two-space pointwise. This all adds up to the fact that if $2 \neq |\sigma|$ divides $q + 1$ then either σ is a homology or σ is fixed point free.

Suppose that $2 \neq |\sigma|$ divides $q - 1$. Then σ leaves at least two lines through the origin invariant and on each of these there are at least two invariant 1-spaces.

In this case if σ has an invariant 1-space which it does not fix pointwise, there is a scalar λ such that $\sigma\lambda$ does fix the one-space Q pointwise, the 1-space in question. If σ leaves all 1-spaces on some line invariant, either σ is a homology or there is a scalar λ such that $\sigma\lambda$ is a homology.

Now let us go back to looking at things more generally. Suppose that σ is an element acting on a vector space of dimension t over GF(q) and suppose that σ has an irreducible constituent V_1 of dimension s. Then $\langle\sigma\rangle$ must be fixed point free on V_1 and $|\sigma|$ must divide $q^s - 1$. Now suppose that $|\sigma|$ is a prime factor of $q^t - 1$ but does not divide $q^s - 1$ for any $s < t$. Then $\langle\sigma\rangle$ must act irreducibly on the vector space of dimension t. We say that $|\sigma|$ is a q-primitive divisor of $q^t - 1$. (This notion can also be defined for numbers which are not prime.)

Now an odd prime factor of $q^2 + 1$ is a q-primitive divisor of $q^4 - 1$.

Hering [9] has investigated subgroups of GL(t, q) which contain q-primitive factors of $q^t - 1$. (His results are stated for the case where q is a prime. If I am not mistaken, they still hold if q is a prime power.)

In Hering's notation, let r be a prime q-primitive factor of $q^n - 1$ and let G be some subgroup of GL(n, q) whose order is divisible by r. Let S be the normal subgroup generated by all r-elements and let F be the Fitting subgroup of G. Then SF/F is simple. If this factor group is trivial then the Sylow r-subgroup R must be in F. Since F is nilpotent, this implies that G has a normal abelian irreducible subgroup and is isomorphic to a subgroup of $\Gamma L(1, q^n)$.

Another result in the same paper states that F is in the centralizer in G of S except possibly when $r = n + 1 = 2^a + 1$ for some a. We want to use this to examine the possibility that SF/F is a cyclic group of prime order. Suppose that $F \cap S$ is in the centre of S and is not fixed point free. Then the subspace pointwise fixed by any element of $F \cap S$ is invariant under S contrary to the fact that S contains elements of order r which must be irreducible. Hence we may assume that $F \cap S$ is f.p.f. Furthermore, if r divides the order of $F \cap S$ we may again conclude that G is isomorphic to a subgroup of $\Gamma L(1, q^n)$.

Thus if $SF/F = S/S \cap F$ has prime order and $G \not\cong \Gamma L(1, q^n)$ either $r = n + 1 = 2^a + 1$ or $F \cap S$ is f.p.f. and its order is not divisible by r. In this case we must have $|S/S \cap F| = r$. We claim that S must be fixed point free. If not, the rth power of an element of S which is not f.p.f. must be in $F \cap S$ and hence must be f.p.f. Hence the non-f.p.f. elements must have order r. But this is impossible if r is a q-primitive divisor. We conclude that if SF/F is a solvable simple group at least one of the following must hold:

(a) $r = n + 1 = 2^a + 1$,

(b) $G \cong \Gamma L(1, q^n)$,

(c) S is fixed point free.

Fixed-point-free linear groups are Frobenius complements and their structure is well known.

When r is not merely a q-primitive divisor of $q^n - 1$ but is a p-primitive divisor, Hering has shown that SF/F cannot be one of the known simple groups of Chevalley type except possibly when S is essentially all of SL(n, q) or r divides (n + 1)(2n + 1) [9, 10].

For n = 4, corresponding to planes of dimension two, the exceptional values of r are 3 and 5. A p-primitive divisor must divide $q^2 + 1$ and 3 does not do so for any q.

For completely reducible solvable groups or for groups whose order is prime to the characteristic there are theorems due respectively to Ito and Feit-Thompson which require Sylow subgroups to be normal for primes too large compared to the dimension. See Dixon [1].

REFERENCES

Note that further references are given in [11], [14], and [20]

1. J.D. Dixon. The structure of linear groups. Van Nostrand Reinhardt (1971).

2. W. Feit. The current situation in the theory of finite simple groups. Actes de Congrès International des Math. 1 (1970): 55-93.
3. D.A. Foulser. Baer p-elements in translation planes. J. Alg. (to appear).
4. M.E. Harris and C. Hering. On the smallest degree of projective representations of the groups PSL(n, q). Canad. J. Math. XXIII (1971): 90-102.
5. C. Hering. Eine nicht-desarguesche Zwei-fach transitive affine Ebene der Ordnung 27. Abh. Math. Sem. Hamb. 34 (1969): 203-208.
6. —— A new class of quasifields. Math Z. 118 (1970): 56-57.
7. —— On shears of translation planes. Abh. Math. Sem. Hamb. 37 (1972): 258-68.
8. —— On 2-groups operating on projective planes. Ill. J. Math. 16 (1972): 581-595.
9. —— Transitive linear groups and linear groups which contain an irreducible subgroup of prime order. Proc. Conf. on Proj. Planes. Wash. State Univ. Press, Pullman (1973).
11. T.G. Ostrom. Vector spaces and construction of finite projective planes. Arch. Math. 19 (1968): 1-25.
12. —— A characterization of generalized André planes. Math. Z. 110 (1969): 1-9.
13. —— Linear transformations and collineations of translation. J. Alg. 14 (1970): 405-416.
14. —— Finite translation planes, Lecture Notes in Math. No. 158. Springer-Verlag.
15. —— A class of planes admitting elations which are not translations. Arch. Math. XXI (1970): 214-217.
16. —— Homologies in translation planes. Proc. London Math. Soc. XXVI (1972): 605-629.
17. —— Normal subgroups of collineation groups of finite translation planes. Geom. Ded. 2 (1974): 467-483.
18. —— Elations in finite translation planes of characteristic 3. Abh. Math. Sem. Hamb. (to appear shortly). Note: The condition that the group is generated by elations was left out in the statement of the main theorem at the beginning of the paper. This condition is definitely a part of the hypothesis.
19. —— Non-modular collineation groups of translation planes. Submitted to Geom. Ded.
20. —— Classification of finite translation planes. Proc. Conf. on Proj. Planes. Wash. State Univ. Press, Pullman, WA (1973).
21. D.S. Passman. Finite Permutation Groups. W.A. Benjamin Pub. Co. (1965).
22. M.L.N. Rao, J.W. Rodabaugh, F.W. Wilke, and J.Z. Zemmer. A new class of finite translation planes ob-

tained from exceptional nearfields. J. Comb. Thy. 11 (1971): 72-92.

Note: This research was supported in part by the National Science Foundation of the U.S.A.

RECENT RESULTS ON PRE-HJELMSLEV GROUPS

Edzard Salow

Let G be a group generated by an invariant set S of involutions. Let P be a non-empty invariant subset of $\{x;\ x \in SS \wedge x \text{ involutory}\}$ with $P \cap S = \phi$. The elements of S are called lines, those of P points. A relation $\mid$ is defined in the following manner:

if $A \in P$, $b \in S$, then

$$A \mid b : \leftrightarrow Ab \text{ is involutory,}$$

if $c, d \in S$ then

$$c \mid d : \leftrightarrow cd \in P.$$

In the first case one says "A is incident with b" and in the second case "c is orthogonal to d." The triple (G, S, P) is called a Pre-Hjelmslev group if

A1. $\forall A \in P, b \in S\ \exists c \in S : c \mid A, b.$

A2. $\forall A \in P, b, c, c' \in S : (A, b \mid c, c' \rightarrow c = c').$

A3. $\forall A, B, C \in P, d \in S : (A, B, C \mid d \rightarrow ABC \in P).$

A4. $\forall a, b, c \in S, D \in P : (a, b, c \mid D \rightarrow abc \in S).$

(G, S, P) is called a (non-elliptic) Hjelmslev group if

in addition $P = \{x;\ x \in SS \wedge x \text{ involutory}\}$.

In this paper we only consider Pre-Hjelmslev groups (G, S, P) which have the following property, called Axiom W:

$\exists a, b, c, d \in S: a \mid b \wedge c \mid d \wedge a, b, c, d$ pairwise intersect uniquely.

This property is equivalent to the existence of $\alpha \in G'$ with $|C_P(\alpha)| = 1$.

THEOREM 1. _Let_ (G, S, P) _be a finite Pre-Hjelmslev group with Axiom W and_ $C_G(P) = \{1\}$. _Then there exist_ R, M, ξ, ℓ, k, ϕ _with the following properties_:

(1) M _is a finite unitary module over the commutative ring_ R _with_ 1 _and_ $\frac{1}{2}$. $xM = \{0\}$ _implies_ $x = 0$ _for_ $x \in R$.

(2) ξ _is an embedding of the incidence structure_ $(P, S, |)$ _into the incidence structure_ $(M \times M, \{R(u, v, w);\ u, v \in R \wedge w \in M \wedge (\exists c, d \in R: uc + vd = 1)\}, I)$. _Here_ $R(u, v, w)\ I\ (a, b)$ _is defined by_ $ua + vb + w = 0$. $R(u, v, w) \in S\xi$ _is equivalent to_ $R(u, v, 0) \in S\xi$. $P\xi = M \times M$.

(3) k _and_ $k + 1$ _are units in_ R. ℓ _is a symmetrical bilinear form on_ M _such that_ $(x, y)\ell z = (x, z)\ell y$ _for all_ $x, y, z \in M$. _The values of_ ℓ _are elements of the radical of_ R. _A symmetrical bilinear form_ f _on_ $R \times R \times M$ _is defined by_ $((u, v, w), (u', v', w'))f := uu'k + vv' + (w, w')\ell$. _If_ $g_1, g_2 \in S$ _and_ $g_i\xi = Rq_i$, _then_ $g_1 \mid g_2$ _is equivalent to_ $(q_1, q_2)f = 0$. _If_ $g \in S$ _and_ $g\xi = Rq$ _then_ $(q, q)f$ _is a unit in_ R.

(4) ϕ *is a monomorphism from* G *into the orthogonal group of the metrical module* ($R \times R \times M$, f). $g\phi$ *is a symmetry along* $g\xi$ *for all* $g \in S$.

By a symmetry along Rc ($c \in R \times R \times M$ and (c, c)f unit in R) we mean the mapping

$\sigma_{Rc} : x \mapsto x - 2(c, x)f((c, c)f)^{-1}c$ for $x \in R \times R \times M$.

THEOREM 2. *Given* R, M, k, ℓ, f *with the properties* (1), (3) *in Theorem 1*. Let $\bar{S} := \{\sigma_{Rq};\ q \in R \times R \times M \wedge (q, q)f$ unit$\}$ *and* $\bar{P} := \{\sigma_{Rq}\sigma_{Rr};\ \sigma_{Rq}, \sigma_{Rr} \in \bar{S} \wedge (q, r)f = 0\}$. *Then* $(<\bar{S}>, \bar{S}, \bar{P})$ *is a finite Pre-Hjelmslev group with Axiom W and* $C_{<\bar{S}>}(\bar{P}) = \{1\}$. $(<\bar{S}>, \bar{S}, \bar{P})$ *is a Hjelmslev group if and only if* 0 *and* 1 *are the only idempotent elements in* R.

It is possible to formulate Theorem 1 in a more group-theoretical manner. R. Stölting has given a characterization of the groups of finite Hjelmslev groups by group-theoretical properties [6]. A similar statement is the following:

G is the group of a finite Pre-Hjelmslev group

if and only if

(*) G has a normal 2-complement K. A Sylow 2-group of G has a generating set T of involutions with TTT = T. There exists an involution $A \in TT$ such that $C_K(\{A, b\}) = \{1\}$ for all $b \in T$.

Now we can reformulate Theorem 1 in the following manner.

THEOREM 3. Let G be a finite group with the properties (*) such that there exists $\alpha \in G'$ with $C_K(A) = C_K(\alpha)$. Then $|G/C_G(G')| \leq 2$ or $G/C_G(G')$ is isomorphic to a subgroup of the little orthogonal group H of a metrical module, which contains H".

By the little orthogonal group we mean the group generated by the set of all symmetries.

The difficulties in the proof of Theorem 1, which is yet unpublished, arise from the fact that there are some strange phenomena which can occur in the finite Pre-Hjelmslev groups with Axiom W: There may exist points A, B with no or with more than one joining line. There may exist a line g with the property that there are no points A, B on g, for which all joining lines carry the same points. There may exist lines a, b, c with $a \mid b$ such that c intersects a and b both in more than one point. This last phenomenon cannot occur in finite Hjelmslev groups.

Among the Pre-Hjelmslev groups which are treated in Theorem 1, those which are singular, i.e. for which PPP = P, play a distinguished role. The bilinear form ℓ in Theorem 1 belonging to a singular Pre-Hjelmslev group is zero. If in proving Theorem 1 one would follow the usual way, one would try to coordinatize the Pre-Hjelmslev group by first coordinatizing a suitable incidence structure. But there is a better way, namely to replace the incidence structure by a singular Pre-Hjelmslev group. This makes good sense, because a singular Pre-Hjelmslev group can be coordinatized directly by half-rotations [2]. The reduction to the singular case is explained in the following proposition.

PROPOSITION. Let (G, S, P) be a finite Pre-Hjelmslev group with Axiom W and $Z(G) = \{1\}$. Let $O \in P$ and m be the mapping from $S \times S$ to the set of rotations around O with $(g, h)m = (O, g)(O, h)$ for $g, h \in S$. ((O, g) denotes the perpendicular on g through O). Then for all $f \in S$ there exists an automorphism κ_f of (P, S, |) which fixes f and the points on f and for which $(g, h)m = ((g\kappa_f, h\kappa_f)m)^{-1}$ for all $g, h \in S$. Define $T := \{\kappa_f; f \in S\}$ and $Q := \{\kappa_e\kappa_f; e, f \in S \wedge (e, f)m = \mathrm{O}\}$. (<T>, T, Q) is a finite singular Pre-Hjelmslev group with Axiom W. The mapping $\tau: h \in T \mapsto \kappa_h, A \in Q \mapsto \kappa_e\kappa_f$ (e, f lines with e, f | A and $(e, f)m = \mathrm{O}$) is an isomorphism from (P, S, |) on (Q, T, |). $\kappa_g \mid \kappa_h$ is equivalent to $(g, h)m = \mathrm{O}$.

For infinite Pre-Hjelmslev groups there doesn't exist such a general coordinatization theorem as Theorem 1. But in [3] we show that an infinite Pre-Hjelmslev group (G, S, P) with Axiom W can be embedded into an orthogonal group over a commutative ring R with 1 and $\frac{1}{2}$ if (G, S, P) has the following two additional properties: (1) All points A, B have a joining line. (2) Every line g is incident with points A, B, which have g as unique joining line. Here

$(P, S, |)$ is embedded into the incidence structure (M, N, I) with $M := \{Rx;$ x element of a basis of $R \times R \times R\}$, $N := \{Rf;$ f element of a basis of $(R \times R \times R)^*\}$ and $I := \{(Rx, Rf); xf = 0\}$. The embedding is locally surjective, i.e. if a point C is an image, then every line through C is an image too. For an arbitrary commutative ring R with 1 and $\frac{1}{2}$ (M, N, I) generally doesn't satisfy property (1). Therefore not every commutative ring with 1 and $\frac{1}{2}$ can occur. That is the reason why it would be more satisfactory to take a weaker axiom about joining lines instead of (1) and (2), for example the following one:

AXIOM Y. There exist $A \in P$, $b \in S$ such that for every point C on b there is a line through A which intersects b uniquely in C.

But it is not known how to coordinatize an infinite Pre-Hjelmslev group with Axiom W and Axiom Y. Nevertheless in [4] and [5] the assumption of Axiom W and Axiom Y has brought some deeper results.

At first we only assume Axiom W and $|P| > 1$ for the Pre-Hjelmslev group (G, S, P). Let Δ be a set of rotations with $C_P(\Delta) \neq \phi$. (By a rotation we mean a product of two intersecting lines.) $C_P(\Delta)$ is the point set of a Sub-pre-Hjelmslev group [1]. If $A \in P$ we denote $\{C_P(\Delta);\ \Delta$ a set of rotations with $A \in C_P(\Delta)\}$ by $\mathfrak{F}_A$. $\mathfrak{F}_A$ becomes a complete lattice by inclusion. Let $K \in \mathfrak{F}_A$. Then there is exactly one partition [K] of P which has K as an element and which is induced by a homomorphism from (G, S, P) on a Pre-Hjelmslev group. Now we consider $C_G([K])$. $C_G([K])$ is a normal subgroup of G. The factorization by $C_G([K])$ gives a homomorphism on a Pre-Hjelmslev group, which identifies exactly the elements of K with A. Let π_A be the mapping from $\mathfrak{F}_A$ into itself with $L\pi_A := C_P(C_G([L]) \cap C_G(A))$ for $L \in \mathfrak{F}_A$. Then π_A is an antihomomorphism from the lattice $\mathfrak{F}_A$ into itself with $\pi_A^3 = \pi_A$. If in addition (G, S, P) satisfies Axiom Y, then π_A^2 is the identity of $\mathfrak{F}_A$. This is the statement of Hjelmslev's Reciprocity Theorem. If $B \in P$ then for every $K \in \mathfrak{F}_A$ there exists an $L \in \mathfrak{F}_B$ with [K] = [L]. Consequently the set $\mathfrak{F} := \{[K]; K \in \mathfrak{F}_A\}$ is independent of the point A. The order relation of $\mathfrak{F}_A$ induces an order relation on $\mathfrak{F}$ such that $(\mathfrak{F}_A, \subseteq)$ is isomorphic to $(\mathfrak{F}, \leq)$ for all $A \in P$. The mapping π, which is induced on $\mathfrak{F}$ by π_A, is independent of the point A too.

The mapping π is interesting in several respects. Again let $A \in P$ and $K \in \mathfrak{F}_A$. Then there exists a locally surjective embedding of the incidence structure of the Pre-Hjelmslev group $(G, S, P)/C_G([K])$ into the incidence structure of $K\pi_A$. (The point set of the incidence structure of $K\pi_A$ is $K\pi_A$ itself, and m is a line if there exists $b \in S$ with $m = \{X \in K\pi_A; X \mid b\}$.) After having got this embedding, one can easily prove for a Pre-Hjelmslev group (G, S, P) with Axiom W and Axiom Y: If U is an antiatom of the lattice $\mathcal{F}$, then $(G, S, P)/C_G(U)$ is a Hjelmslev group with Axiom W and Axiom Y, in which no double incidences occur, i.e. no lines a, b intersect in more than one point.

Finally we want to consider more general partitions of P. A homomorphism which is induced by an element of $\mathcal{F}$ we call an elementary contraction. A product of such elementary contractions we call a contraction. There are contractions which are not elementary contractions. The set of all partitions of P which are induced by contractions we denote by $\mathcal{E}$. On $\mathcal{E}$ there exists exactly one associative operation $\circ$ with the following property: If $U \in \mathcal{E}$, $[C_P(\Delta)] \in \mathcal{F}$, ϕ is the homomorphism from (G, S, P) induced by U, and ψ is the homomorphism from $(G, S, P)\phi$ induced by $[C_P(\Delta\phi)]$, then $U \circ [C_P(\Delta)]$ is the partition of P which is induced by $\phi\psi$. It is an astonishing fact that $\circ$ is commutative. $\mathcal{F}$ generates $\mathcal{E}$. Now we assume that (G, S, P) is a Pre-Hjelmslev group with Axiom W and Axiom Y, such that the ascending chain condition is valid for $\mathcal{F}$. Then if $\{U_i\}_{i \in \mathbb{N}}$ is a sequence of elements of $\mathcal{F}$, $\{U_1 \circ \ldots \circ U_n\}_{n \in \mathbb{N}}$ is finite. It follows that $\mathcal{F}$ has only a finite number of atoms and antiatoms. The structure of (G, S, P) can be clarified even more. There is a finite chain $U_0 < U_1 < \ldots < U_n$ of elements of $\mathcal{E}$ with $U_0 = \{\{A\}; A \in P\}$ and $U_n = \{P\}$ which has the following property: Let $A \in P$ and $K_i \in U_i$ with $A \in K_i$. K_i is the point set of a Sub-pre-Hjelmslev group H_i of (G, S, P) and there exists a homomorphism from

H_i onto a Hjelmslev group without double incidences which identifies exactly the points of K_{i-1} with A.

REFERENCES

1. F. Bachmann. Hjelmslev-Gruppen. Mathematisches Seminar der Universität Kiel 1970-71.
2. E. Salow. Singuläre Hjelmslev-Gruppen. Geom. Dedicata 1 (1973): 447-467.
3. —— Einbettung von Hjelmslev-Gruppen in orthogonale Gruppen über kommutativen Ringen. Math. Z. 134 (1973): 143-170.
4. —— Fixpunktmengen von Drehungen in Hjelmslev-Gruppen. Abh. Math. Sem. Univ. Hamburg 41 (1974): 37-63.
5. —— Ketten homogener Fleck-Überdeckungen von Prä-Hjelmslev-Gruppen. Abh. Math. Sem. Univ. Hamburg 41 (1974): 64-71.
6. R. Stölting. Über endliche Hjelmslev-Gruppen. Math. Z. 135 (1974): 249-256.

FINITE GEOMETRIC CONFIGURATIONS[1]

J.J. Seidel

1. INTRODUCTION

We present five elementary examples in 3- and 4-dimensional spaces over Galois fields GF(q), over the reals $\mathbb{R}$, and over the complexes $\mathbb{C}$. They serve as an introduction for the subsequent sections. These sections treat discrete geometry, that is a mixture (with a geometric flavour) of discrete mathematics, combinatorics, finite groups, geometry which has certain applications in coding theory, statistical designs, network theory, and graph theory.

1.1. Hadamard matrices

Given a cube in $\mathbb{R}^3$ (with side 2, centre at the origin, sides parallel to the coordinate axes), we wish to construct a regular tetrahedron by selecting 4 among the 8 vertices of the cube. The answer is provided by the points

$$\begin{matrix}(1,\ 1,\ 1)\\(1,-1,-1)\\(-1,1,-1)\\(-1,-1,1)\end{matrix}\qquad \text{The matrix } \begin{bmatrix}1&1&1&1\\1&1&-1&-1\\1&-1&1&-1\\1&-1&-1&1\end{bmatrix} =: H_4$$

1. Notes taken by J.G. Sunday.

is a Hadamard matrix of order 4.

DEFINITION. A _Hadamard matrix_ H_r of the order r is a square matrix of order r, with elements ±1, satisfying

$$H_r H_r^T = rI_r.$$

Necessary conditions for the existence of H_r are

$$r = 2, \quad r \equiv 0 \pmod 4.$$

It has been conjectured that these conditions are sufficient. This conjecture has been verified for r < 188 and for several infinite series. For the state of affairs we refer to M. Hall [6], J.H. van Lint [18], J.S. Wallis [20].

1.2. A perfect ternary code

In the vector space V(4, 3) of dimension 4 over GF(3) consider the plane spanned by

$$f = (1, 0, 1, 2), \quad g = (0, 1, 1, 1).$$

There are 9 vectors in this plane, all but the origin having 3 non-zero coordinates. Hence each pair of the 9 vectors has 3 coordinates in which they differ (has _Hamming distance_ 3). Calling _words_ the 81 vectors in V(4, 3), _code words_ the 9 vectors in the plane, we have found a 1-error-correcting code. This _code_ (the plane) is linear and _perfect_. This means the following. Around each code word we draw a sphere of (Hamming) radius 1, containing 8 words. The 9 spheres are mutually disjoint, and exhaust V(4, 3) by the count

$$9(1 + 8) = 81.$$

1.3. The binary projective 3-space PG(3, 2)

Consider the vector space V(4, 2) of dimension 4 over GF(2) provided with the non-degenerate alternating bilinear form

$$B(x, y) = \xi_1\eta_2 + \xi_2\eta_1 + \xi_3\eta_4 + \xi_4\eta_3,$$

for $x = (\xi_1, \xi_2, \xi_3, \xi_4)$, $y = (\eta_1, \eta_2, \eta_3, \eta_4)$.

We make the following observations:

1. The symplectic group Sp(4, 2) acts transitively on $V(4, 2) \setminus \{0\}$.

2. The 16×16 matrix $[B(x, y)]_{x,y \in V}$ yields $[2B - J]$, which is a Hadamard matrix of order 16 (here J is the all one matrix of order 16).

3. The columns of the matrix

$$[B \quad J - B]$$

constitute a binary (32, 16, 8)-code.

4. Consider the graph whose vertices are the 15 vectors of $V \setminus \{0\}$, in which x and y are adjacent if and only if $B(x, y) = 0$. This <u>symplectic graph</u> is strongly regular, and even rank 3, under the action of Sp(4, 2). Obviously, Sp(4, 2) acts <u>not</u> 2-transitively on the graph!

5. The extension of Sp(4, 2) by the translations of V(4, 2) does act 2-transitively on the vectors of V(4, 2). It maps the set of the triples $\{x, y, z\}$ with

$$B(x, y) + B(y, z) + B(x, z) = 0$$

onto itself. This set defines a <u>regular two-graph</u>.

6. Since $Sp(4, 2) \simeq Symm(6)$ the graph under 4 is easily drawn.

1.4. Equiangular lines

In a set of equiangular lines the angle ϕ between <u>each</u> pair of distinct lines is the same. We consider two examples in $\mathbb{R}^3$, where we have chosen a unit vector p_i along each line (in either of the 2 directions), and their matrix C defined by

$$P = [\langle p_i, p_j \rangle], \quad C = \frac{1}{\cos\phi}[P - I].$$

The examples are the 4 diagonals of the cube and the 6 diagonals of the icosahedron:

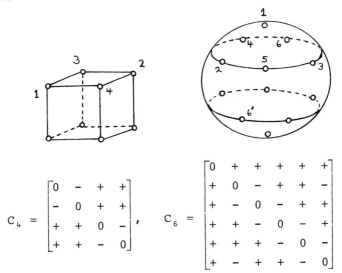

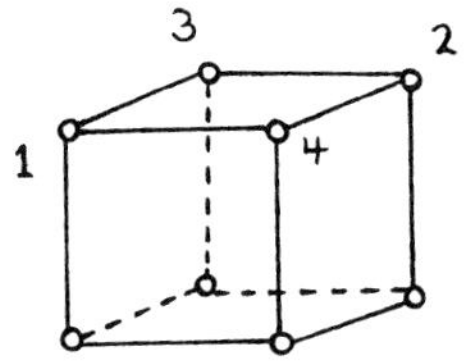

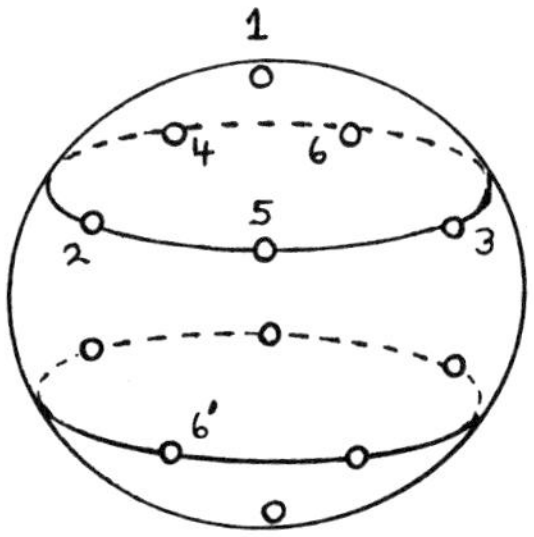

$$C_4 = \begin{bmatrix} 0 & - & + & + \\ - & 0 & + & + \\ + & + & 0 & - \\ + & + & - & 0 \end{bmatrix}, \quad C_6 = \begin{bmatrix} 0 & + & + & + & + & + \\ + & 0 & - & + & + & - \\ + & - & 0 & - & + & + \\ + & + & - & 0 & - & + \\ + & + & + & - & 0 & - \\ + & - & + & + & - & 0 \end{bmatrix}$$

We observe that

$$(C_4 - I)(C_4 + 3I) = 0, \qquad C_6^2 = 5I,$$

$$\cos\phi = \frac{1}{3}, \qquad \cos\phi = \frac{1}{\sqrt{5}}.$$

With these examples, we associate the graphs

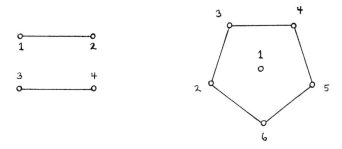

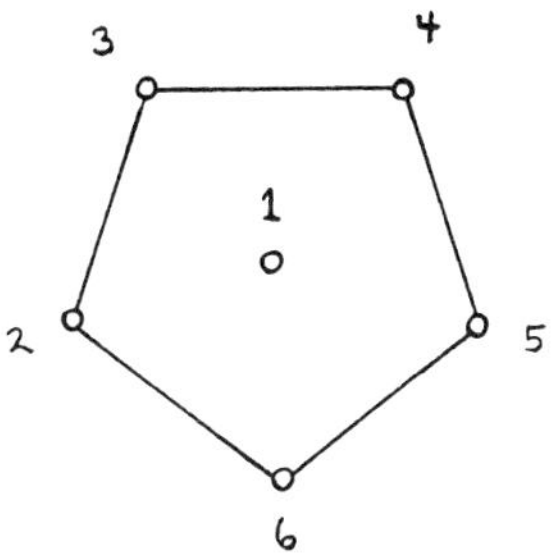

If, in the second example, we take 6' instead of 6, then the graph is switched into

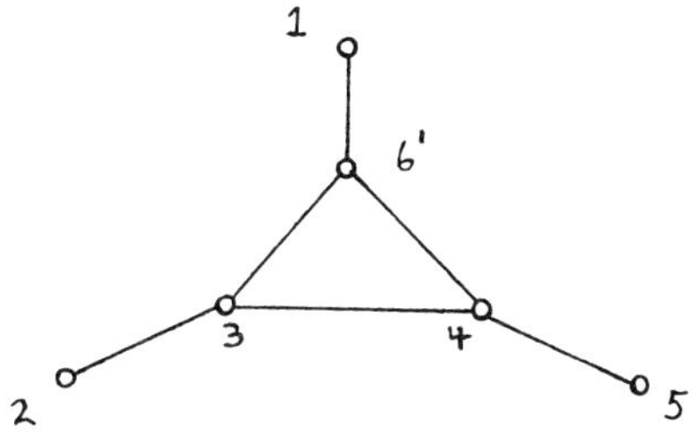

The effect on the matrix C_6 of this new choice of 6' is

$$C' = DCD$$

where D is a diagonal matrix, whose diagonal elements are 1, 1, 1, 1, 1, -1. This exemplifies the relations between sets of equiangular lines, classes of graphs, and <u>two-graphs</u>. Indeed, the triples of lines are of two kinds: those with $(p_h, p_i)(p_i, p_j)(p_j, p_h) > 0$ such as (h, i, j) = (1, 4, 6), and those with $(p_h, p_i)(p_i, p_j)(p_j, p_h) < 0$ such as (h, i, j) = (1, 2, 3). In the second example, the automorphism group of the set of equiangular lines is the icosahedral group A_5. It contains as subgroups the dihedral groups D_{10} and D_6, the automorphism groups of the two graphs mentioned above, which are related by switching with respect to vertex 6.

1.5. Complex equiangular lines

We rearrange the Hadamard matrix H_4 of example 1.1 into

$$\begin{bmatrix} + & + & + & + \\ - & + & - & + \\ - & + & + & - \\ - & - & + & + \end{bmatrix} = I + \begin{bmatrix} 0 & + & + & + \\ - & 0 & - & + \\ - & + & 0 & - \\ - & - & + & 0 \end{bmatrix} = I + C.$$

The skew matrix C, multiplied by the complex number i, is a hermitian matrix satisfying

$$(iC)^2 = 3I.$$

Hence $iC + I\sqrt{3}$ is hermitian positive semidefinite of rank 2, and may be considered as the matrix of the hermitian products of the vectors $(\sqrt{3}, 0)$, $(i, \sqrt{2})$, $(i, \omega\sqrt{2})$, $(i, \omega^2\sqrt{2})$ where ω is a primitive cube root of unity. Hence we have 4 equiangular lines in the complex 2-dimensional plane $\mathbb{C}^2$. In real space $\mathbb{R}^4$ these correspond to 4 equi-isoclinic planes.

2. CODES

2.1. Definitions

Using a finite set S of cardinality s as an alphabet, we may form s^n distinct words of length n; that is, each element of the set

$$S^n = \{\underline{x} = x_1x_2\ldots x_n \mid x_i \in S,\ 1 \le i \le n\}$$

is considered a word. The degree of a "spelling error"

might be measured by the number of positions in which a given word differs from the intended word, so a useful device is the Hamming distance $d(\underline{x}, \underline{y})$ between two words in S^n:

$$d(\underline{x}, \underline{y}) = \text{the number of positions in which } \underline{x} \text{ and } \underline{y} \text{ differ.}$$

It is elementary to verify that d is a metric on S^n, and we may occasionally refer to the closed ball

$$B[\underline{c}; r] = \{\underline{x} \in S^n \mid d(\underline{c}, \underline{x}) \leq r\}.$$

For instance, if $S = \{a, b\}$, then the two words $\underline{x}$ = aabbb and $\underline{y}$ = abaab in S^5 have Hamming distance 3, and B[aabbb; 1] consists of the words aabbb, babbb, abbbb, aaabb, aabab, and aabba.

Often, we want only a subset of S^n to form a vocabulary of meaningful words; such a subset is said to be a code, and its elements are called code words. A code is called an (M, n, d)-code if the code consists of M code words, each of length n, such that the Hamming distance between any two code words is at least d. In this case, if $\underline{c}$ is a code word, then any element of $B\left[\underline{c}; [\frac{d-1}{2}]\right]$ is closer to $\underline{c}$ than to any other code word, so a word differing from $\underline{c}$ in at most $[\frac{d-1}{2}]$ places might reasonably be assumed to be $\underline{c}$, up to a spelling error; we say that the code is $[\frac{d-1}{2}]$-error-correcting. If, moreover, the family

$$\left\{B\left[\underline{c}; [\tfrac{d-1}{2}]\right] \;\middle|\; \underline{c} \text{ is a code word}\right\}$$

of mutually disjoint closed balls covers all of S^n, we say that the code is perfect.

2.2. Binary codes

For a binary code, one may easily take the alphabet S to be {1, -1}. Abbreviating 1 and -1 to + and - respectively, we may group corresponding positions for two words $\underline{x}$, $\underline{y}$ in S^n as follows

$\underline{x}$	+ + + +	+ + + +	- - - -	- - - -
$\underline{y}$	+ + + +	- - - -	+ + + +	- - - -
No. of positions	P_1	P_2	P_3	P_4

(where, of course, $P_1 + P_2 + P_3 + P_4 = n$), so that the Hamming distance is given by

$$d(\underline{x}, \underline{y}) = P_2 + P_3$$

and the usual inner product $\langle\ ,\ \rangle$ is given by

$$(2.A)\quad \langle\underline{x}, \underline{y}\rangle = P_1 - P_2 - P_3 + P_4 = n - 2\cdot d(\underline{x}, \underline{y}).$$

We may use these notions to establish

THE PLOTKIN BOUND. *If the parameters of a binary* (M, n, d) *code satisfy* $n < 2d$, *then* $M \le 2d/(2d - n)$.

PROOF. The M code words may be used to form the rows of an $M \times n$ matrix P. According to (2.A), the Hamming distance between distinct code words is at least d precisely when

$$\langle\underline{x}, \underline{y}\rangle \le n - 2d$$

for all distinct $\underline{x}, \underline{y} \in S^n$. But the inner products are given as entries in the symmetric $M \times M$ matrix PP^t. There fore, for an (M, n, d)-code,

$$(2.B) \quad PP^t \underset{\text{elementwise}}{\leq} nI + (n - 2d)(J - I)$$

$$= 2dI + (n - 2d)J.$$

If j_M is a column vector of M 1's, then $(j_M^t P)(P^t j_M)$ is the square of the ordinary euclidean norm of $j_M^t P$, so (2.B) implies that

$$0 \leq j_M^t PP^t j_M \leq 2dM - (2d - n)M^2.$$

If $n < 2d$, this establishes the Plotkin bound.

Now, a code is not essentially changed if we multiply each entry in a column of P by -1. Thus, it may be assumed that the first row of P consists entirely of 1's, so we may write

$$P = \begin{bmatrix} j_n^t \\ Q \end{bmatrix}$$

where Q is an $(M - 1) \times n$ matrix of 1's and -1's. In the extremal case $M = 2d/(2d - n)$, we find that

$$QQ^t = 2dI - (2d - n)J,$$

$$j_{M-1}^t Q = -j_n^t,$$

$$Qj_n = (n - 2d)j_{M-1},$$

so Q may be regarded as the (+, -)-incidence matrix of a block design.

There are several interesting examples and applications of binary codes. For instance, in the extreme case in which the Plotkin bound is valid (i.e. $2d - n = 1$ and $M = 2d$), the matrix $[j_M \ P]$ is found to be a Hadamard matrix of order M. If we remove the restriction that led to the Plotkin bound, we may find codes with a larger vocabulary (i.e. more code words). For instance, there is a class of binary (8t, 4t, 2t)-codes obtained from matrices

$$P = \begin{bmatrix} H \\ -H \end{bmatrix}$$

where H is a Hadamard matrix of order 4t. A final example is the binary (16, 5, 2)-code whose code words consist of all vectors $(x_1, x_2, x_3, x_4, x_5)$ with an even number of -1's.

For further details, we refer to [4] and to [15]. The general algebraic theory of codes is exposed in [17].

2.3. Ternary Golay codes

For ternary codes, the alphabet S consists of the elements 0, 1, -1 of the finite field GF(3). In this way, the words of length n may be regarded as forming the vector space V(n, 3). We are particularly interested in situations where the code words form a linear subspace (such a code is called a linear code).

If 1 and -1 are abbreviated to + and - respectively, consider the 6 × 12 matrix

$$[C_6 I_6] = \begin{bmatrix} 0 & + & + & + & + & + & + & 0 & 0 & 0 & 0 & 0 \\ + & 0 & - & + & + & - & 0 & + & 0 & 0 & 0 & 0 \\ + & - & 0 & - & + & + & 0 & 0 & + & 0 & 0 & 0 \\ + & + & - & 0 & - & + & 0 & 0 & 0 & + & 0 & 0 \\ + & + & + & - & 0 & - & 0 & 0 & 0 & 0 & + & 0 \\ + & - & + & + & - & 0 & 0 & 0 & 0 & 0 & 0 & + \end{bmatrix}$$

(where C_6 is the same matrix as in §1.4). Since the column rank is obviously 6, the rows form a basis for a subspace W of dimension 6 in V(12, 3). Now, the Hamming distance between any pair of these basis vectors is congruent to 0 (mod 3), so any pair of distinct elements from W has Hamming distance 3, 6, 9, or 12. Moreover, it can be shown that no two code words in W have Hamming distance exactly 3. It follows that the 3^6 words in W form a ternary (3^6, 12, 6)-code, called the Extended Golay code. The automorphism group of this code is the Mathieu group M_{12}.

The ordinary Golay code is obtained by deleting any one column of the matrix $[C_6 \ I_6]$. The shortened rows then form a basis for a subspace U of dimension 6 in V(11, 3). The Hamming distance between any pair of distinct words in U is at least 5, so the 3^6 words of length 11 in U form a ternary (3^6, 11, 5)-code. Moreover, each closed ball $B[\underline{c}; 2]$ contains exactly

$$1 + 2\cdot 11 + 2^2\cdot\binom{11}{2} = 3^5$$

words of V(11, 3), so the family

$$\{B[\underline{c}; 2] \mid \underline{c} \in U\}$$

of disjoint closed balls covers $3^6\cdot 3^5$ words of V(11, 3). Since there happen to be exactly 3^{11} words in V(11, 3), we see that this Golay code is an example of a perfect code.

A third ternary code is obtained as the complement of the Golay code. More precisely, if we take the 3^5 words of the completely orthogonal subspace $U^\perp$ in V(11, 3) as code words, we obtain a ternary $(3^5, 11, 6)$-code. The code words of this last code have Hamming distances 6 and 9 only.

REMARK. There are also binary Golay codes with words belonging to $V(n, 2)$: a binary $(2^{12}, 24, 8)$-code which is related to the Mathieu group M_{24}, and a perfect binary $(2^{12}, 23, 7)$-code which is related to M_{23}.

3. STRONGLY REGULAR GRAPHS

3.1. Definitions

Consider a graph G; a vertex P is said to be a <u>neighbour</u> of the (unordered) pair of vertices P_1, P_2 if P is adjacent to both P_1 and P_2.

LEMMA 3.1. <u>Let</u> G <u>be a graph with</u> n <u>vertices such that</u>

(i) <u>each vertex is adjacent to at most</u> k <u>others;</u>

(ii) <u>each adjacent pair of vertices has at least</u> λ <u>neighbours; and</u>

(iii) <u>each non-adjacent pair of vertices has at least</u> μ <u>neighbours.</u>

<u>Then</u> $(n - k - 1)\mu \leq k(k - 1 - \lambda)$.

PROOF. Choose a vertex P_0 of G. Define $A(P_0)$ to be the set of vertices adjacent to P_0, and $NA(P_0)$ to be the set of vertices not adjacent to P_0. By assumption, $|A(P_0)| \leq k$ so $|NA(P_0)| \geq n - 1 - k$. We now proceed to count the number M of edges between $A(P_0)$ and $NA(P_0)$ in two ways. Since each vertex Q in $NA(P_0)$ determines, with P_0, a non-adjacent pair, there are at least μ vertices adjacent to both P_0 and Q; thus

$$M \geq |NA(P_0)|\cdot\mu \geq (n - 1 - k)\mu.$$

Similarly, each vertex R in $A(P_0)$ is adjacent to at most $k - 1$ vertices other than P_0 and has at least λ neighbours with P_0, so there are at most $k - 1 - \lambda$ vertices adjacent to R but not to P_0; hence

$$M \leq |A(P_0)|\cdot(k - 1 - \lambda) \leq k\cdot(k - 1 - \lambda).$$

These two inequalities establish the lemma.

Examining the circumstances under which equality occurs in Lemma 3.1, we say that a graph is <u>strongly regular</u> if

(i) each vertex is adjacent to exactly k others;

(ii) each adjacent pair of vertices has exactly λ neighbours; and

(iii) each non-adjacent pair of vertices has exactly μ neighbours.

For instance, a pentagon is a strongly regular graph with $n = 5$, $k = 2$, $\lambda = 0$, and $\mu = 1$. The Petersen graph

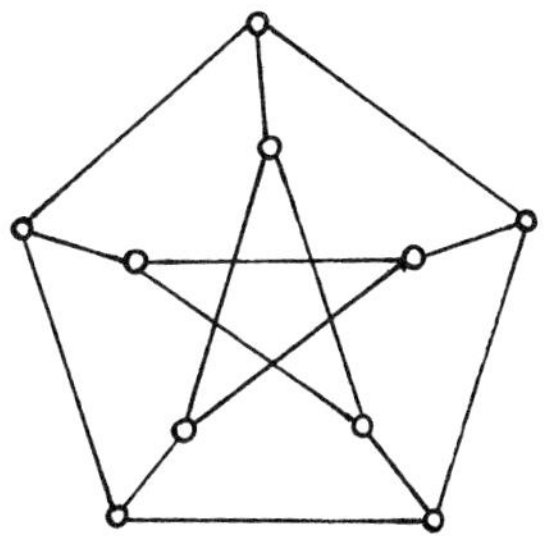

is a strongly regular graph with $n = 10$, $k = 3$, $\lambda = 0$, and $\mu = 1$. Moreover, if T is a finite set of cardinality t, the $\binom{t}{2}$ unordered pairs of distinct elements of T form the vertices of a strongly regular graph (called a triangular graph T(t)) with $n = \frac{1}{2}t(t - 1)$, $k = 2(t - 2)$, $\lambda = t - 2$, and $\mu = 4$; two pairs are called adjacent precisely when they have one element in common.

3.2. Graphs and matrices

The properties of a strongly regular graph are often studied via the corresponding adjacency matrix. In particular, we state without proof

THEOREM 3.2. *Suppose* A *is a symmetric matrix with entries* 0 *and* 1 *such that all diagonal entries are zero*. *Then* A *is the adjacency matrix of a strongly regular graph with parameters* (n, k, λ, μ) *if and only if there are numbers* r *and* s *such that*

$$Aj_n = k \cdot j_n, \quad (A - rI)(A - sI) = \mu J.$$

Moreover, r *and* s *are the roots of* $x^2 + (\mu - \lambda)x + \mu - k = 0$. *In addition*, k, r, *and* s *are the eigenvalues of* A, *the eigenvalue* k *has multiplicity* 1, *and*, *if* $r + s \neq -1$, *the eigenvalues* r *and* s *are both integers*.

The adjacency matrix A_s of a pentagon satisfies $A_5^2 + A_5 - I = J$ so $r + s = -1$; in this instance, both r and s happen to be irrational. Similarly, the adjacency matrix A_{10} of the Petersen graph satisfies $A_{10}^2 + A_{10} - 2I = J$ so

$r + s = -1$ again; in this case, however, r and s happen to be integers.

The state of affairs for the special case in which $r + s = -1$ while r and s are both integers is quite interesting. If x and y are the multiplicities of the eigenvalues r and s respectively, we are able to derive

$$(2r + 1)^2 - 2(2r + 1) - 16(x - y)\mu(2r + 1) = 16\mu^2 - 1$$

as a consequence of the proof of the above theorem. It follows that $2r + 1 \mid 16\mu^2 - 1$. One class of such graphs consists of the Moore graphs, which are restricted to $\mu = 1$ and $\lambda = 0$; since $2r + 1 \mid 15$ for this class, we find that the parameters are limited to

r	s	k	n
1	-2	3	10
2	-3	7	50
7	-8	56	3250

The Petersen graph yields the first possibility, and it is at present unknown whether the third possibility is realized. The second possibility is realized by the Hoffman-Singleton graph [7], which we now describe. The 50 vertices of the graph are given by the 15 points and 35 lines of PG(3, 2). To be able to define adjacency, we require the following fact (cf. [1]): "there is a 1-1 correspondence between the lines of PG(3, 2) and the unordered triples from {a, b, c, d, e, f, g} in such a way that two lines intersect or are skew according as the corresponding triples have an odd or an even number of elements in common." Now, we can state that two points are never considered adjacent, a point and a line are considered adjacent only if they are incident, and two lines are considered adjacent precisely when the corresponding triples are disjoint. It is an elementary exercise to check that this graph does have the stipulated properties.

An alternative class of these graphs, cf. [2], is produced by the restriction $\mu = 2$ and $\lambda = 1$, so that $2r + 1 \mid 63$; the parameters in this instance are limited to

r	1	3	4	10	31
k	4	14	22	112	994
n	9	99	243	6273	49401

At present, it is unknown whether the second, fourth, or fifth possibilities may be realized. The 9 vertices for the first possibility may be taken to be the ordered pairs from {a, b, c}; two ordered pairs are adjacent if they agree in one position. The 243 vertices for the third possibility may be taken to be the elements of V(5, 3); two vectors $\underline{x}$ and $\underline{y}$ are called adjacent if one of the differences $\underline{x} - \underline{y}$ or $\underline{y} - \underline{x}$ is a column of the 5×11 matrix of generators for the complement to the Golay code. In fact, if $\underline{c}_1, \underline{c}_2, \ldots, \underline{c}_{11}$ are the columns of this matrix, then there are 22 vectors $\pm\underline{c}_i$, 220 vectors $\pm\underline{c}_i \pm\underline{c}_j$, and 1 vector $\underline{o}$. Since it was shown in §2.3 that no four columns of this matrix are dependent, it follows that these 243 vector are pairwise distinct, so they exhaust V(5, 3) and may be taken as representatives of the vertices of this graph. Again, one may readily check that the graph does have the required properties.

3.3. Rank 3 graphs

If G is a permutation group on a set Ω, we may define a relation $\sim$ on $\Omega \times \Omega$ by

$$(\alpha, \beta) \sim (\gamma, \delta) \text{ if and only if } \gamma = g(\alpha) \text{ and } \delta = g(\beta)$$
$$\text{for some } g \in G.$$

It is an elementary exercise to show that this relation is an equivalence relation, with equivalence classes of the form

$$\{(g(\alpha), g(\beta)) \mid g \in G\}$$

where $\alpha, \beta \in \Omega$. Obviously, if Δ is an equivalence class, then so is

$$\bar{\Delta} := \{(\beta, \alpha) \mid (\alpha, \beta) \in \Delta\}.$$

If G is transitive on Ω, the diagonal τ of $\Omega \times \Omega$ is an equivalence class. Moreover, if G is 2-transitive, there are exactly two equivalence classes, namely τ and $\Omega \times \Omega \backslash \tau$. The next most interesting situation, in which G is slightly less than 2-transitive, may be defined as follows: "G is called a <u>rank 3 group</u> on Ω if G is transitive on Ω, the order $|G|$ is even, and there are exactly three equivalence classes τ, Δ, and Γ." Using the restriction to even order, one may then show that $\bar{\Delta} = \Delta$ and $\bar{\Gamma} = \Gamma$, so each unordered pair of distinct elements of Ω belongs either to Δ or to Γ, but not both. We may use this observation to define two graphs with the elements of Ω as vertices: two vertices α, β are said to be adjacent in one graph if and only if

$(\alpha, \beta) \in \Delta$, and they are adjacent in the other graph if and only if $(\alpha, \beta) \in \Gamma$. Therefore, two vertices are adjacent in one graph precisely when they are non-adjacent in the other. These two complementary graphs are called rank 3 graphs. Since G is transitive on both of the equivalence classes Δ and Γ, it is transitive on the edges and on the non-edges of each of these graphs. Furthermore, G is also transitive on the vertices, so we find that any rank 3 graph is strongly regular.

We give an example to illustrate these remarks. If T is a finite set of cardinality t $(t \geq 4)$ and Ω consists of the $\binom{t}{2}$ unordered pairs from T, it is clear that the symmetric group S_T on T is a rank 3 group on Ω, since the three equivalence classes are

$$\tau = \{(\omega, \omega) \mid \omega \in \Omega\},$$

$$\Delta = \left\{(\omega, \omega') \;\middle|\; \begin{array}{l}\text{the unordered pairs } \omega \text{ and } \omega' \text{ have no}\\ \text{element in common}\end{array}\right\},$$

$$\Gamma = \left\{(\omega, \omega') \;\middle|\; \begin{array}{l}\text{the unordered pairs } \omega \text{ and } \omega' \text{ have}\\ \text{exactly one element in common}\end{array}\right\}.$$

The strongly regular graph with edges taken from Γ is precisely the triangular graph T(t).

4. QUADRATIC FORMS OVER GF(2)

4.1. Background

There are many applications of bilinear and quadratic forms to codes and strongly regular graphs. In order to demonstrate this, we reproduce some results about forms which may be found with detailed references in Dieudonné's treatise La géométrie des groupes classiques.

An alternating bilinear form on V(2m, 2) is a bilinear map $B: V \times V \to GF(2)$ such that $B(\underline{x}, \underline{x}) = 0$ for every $\underline{x} \in V$; it follows that $B(\underline{x}, \underline{y}) = B(\underline{y}, \underline{x})$ for all $\underline{x}, \underline{y} \in V$. If $B(\underline{x}, \underline{y}) = 0$ for all $\underline{y} \in V$ only when $\underline{x} = \underline{0}$, the form is called non-singular.

Since GF(2) contains only two elements 0 and 1, we may consider the entries of the square matrix

$$B := [B(\underline{x}, \underline{y})]_{\underline{x} \in V,\ \underline{y} \in V}$$

(of order 2^{2m}) as integers. With this interpretation, one may show that the matrix $2B - J$, consisting of 1's and (-1)'s, is a Hadamard matrix. On the other hand, we may obtain an example of a binary $(2^{2m+1}, 2^{2m}, 2^{2m-1})$-code from the columns of the $2^{2m} \times 2^{2m+1}$ matrix $[B\ J-B]$.

Using a non-singular alternating bilinear form B on V, we may define the symplectic graph: the vertices are the points of $V \setminus \{\underline{o}\}$, and two vertices $\underline{x}, \underline{y}$ are called adjacent precisely when $\underline{x} \neq \underline{y}$ and $B(\underline{x}, \underline{y}) = 0$. We find that the symplectic graph is a rank 3 graph under the group

$$Sp(2m, 2) = \{\alpha \in GL(V) \mid B(\underline{x}^{\alpha}, \underline{y}^{\alpha}) = B(\underline{x}, \underline{y}) \text{ for all } \underline{x}, \underline{y} \in V\}.$$

Since $W = \{\underline{x} \in V \mid B(\underline{a}, \underline{x}) = o\}$ is a linear subspace of V and $\underline{a}$ is adjacent to every element of $W \setminus \{\underline{o}, \underline{a}\}$, it follows that the symplectic graph is a regular graph with $k = 2^{2m-1} - 2$ edges through each vertex. If $B(\underline{x}_0, \underline{y}_0) = 0$ and $B(\underline{x}_1, \underline{y}_1) = 1$, then

$$U_0 = \{\underline{z} \in V \mid B(\underline{z}, \underline{x}_0) = 0 = B(\underline{z}, \underline{y}_0)\}$$

and $U_1 = \{\underline{z} \in V \mid B(\underline{z}, \underline{x}_1) = 0 = B(\underline{z}, \underline{y}_1)\}$

are both the intersection of two hyperplanes in $V(2m, 2)$, so they each contain exactly 2^{2m-2} elements; it follows that the symplectic graph is strongly regular, with

$$\lambda = |U_0 \backslash \{\underline{0}, \underline{x}_0, \underline{y}_0\}| = 2^{2m-2} - 3$$

and $\mu = |U_1 \backslash \{\underline{o}\}| = 2^{2m-2} - 1.$

A quadratic form on $V(2m, 2)$ is a map $Q: V \to GF(2)$ such that $Q(\underline{x} + \underline{y}) + Q(\underline{x}) + Q(\underline{y})$ is a bilinear form $B(\underline{x}, \underline{y})$ on V; it follows that $Q(\underline{o}) = 0$ and that the associated bilinear form is alternating. If this bilinear form is also non-singular, it is known that there is a basis $\underline{e}_1, \ldots, \underline{e}_{2m}$ of V with respect to which the quadratic form Q is given by

$$Q(\underline{x}) = Q^+(\underline{x}) = \xi_1\xi_2 + \xi_3\xi_4 + \ldots + \xi_{2m-1}\xi_{2m}$$

or $Q(\underline{x}) = Q^-(\underline{x}) = \xi_1^2 + \xi_1\xi_2 + \xi_2^2 + \xi_3\xi_4 + \ldots + \xi_{2m-1}\xi_{2m}$

and that the number of zeros of Q is $2^{2m-1} + 2^{m-1}$ or $2^{2m-1} - 2^{m-1}$ respectively. Let Ω represent the set of all zeros of a non-singular Q.

We may now define the <u>orthogonal graph</u>: the vertices are the elements of $\Omega \backslash \{\underline{o}\}$ (alternatively, one may use the elements of $V \backslash \Omega$), and two distinct vertices $\underline{x}$, $\underline{y}$ are adjacent precisely when $Q(\underline{x} + \underline{y}) = 0$. As before, we find that the orthogonal graph is a rank 3 graph under the orthogonal group

$$\{\alpha \in GL(V) \mid Q(\underline{x}^{\alpha}) = Q(\underline{x}) \text{ for all } x \in V\}.$$

We mention that the symplectic and orthogonal graphs may be characterized in a more geometric manner (cf. [10], [13]):

SHULT'S THEOREM. _For a regular graph_ G _which is not a complete graph, the following are equivalent_:

(i) G _is isomorphic to_ a _symplectic or an orthogonal graph_;

(ii) _For each pair of adjacent vertices_ a, b _of_ G, _there is a vertex_ c _adjacent to both_ a _and_ b _such that each further vertex is adjacent to an odd number of vertices of the triangle_ {a, b, c}.

4.2. Kerdock sets

We now investigate some related concepts following Kerdock, Patterson, and Goethals. Suppose $\mathcal{B}$ is the set of all alternating bilinear forms on V(2m, 2), allowing for singular forms and even the zero-form. Each form $B(\underline{x}, \underline{y})$ may be regarded as a matrix product

$$\underline{x}^t B \underline{y}$$

where B is a symmetric $2m \times 2m$ matrix with entries in GF(2), such that the main diagonal consists entirely of zeros.

Since there are m(2m - 1) positions in such a matrix above the main diagonal, we find there are exactly $2^{m(2m-1)}$ elements in *B*.

Consider a subset *S* of *B* such that whenever B_i and $B_j \in S$ the form $B_i + B_j$ is non-singular. If the matrices representing B_i and B_j have the same first row, then the matrix representing $B_i + B_j$ has first row 0, and $B_i + B_j$ is singular. Therefore, the first rows for the matrices of the B_i in *S* must all be distinct, so *S* contains at most 2^{2m-1} forms. For this maximal case, we define: a subset *K* of *B* is called a <u>Kerdock set</u> if $|K| = 2^{2m-1}$ and, whenever B_i and B_j are distinct elements of *K*, the form $B_i + B_j$ is non-singular. We give the following construction of a Kerdock set. If V(2m, 2) is considered to be $GF(2^{2m-1}) \oplus GF(2)$, each element $\underline{x}$ in V may be written uniquely as $\underline{x} = \alpha + \xi$ where $\alpha \in GF(2^{2m-1})$ and $\xi \in GF(2)$. The trace of an element $\alpha \in GF(2^{2m-1})$ is defined by

$$Tr(\alpha) = \sum_{k=0}^{2m-2} \alpha^{(2^k)}$$

and it is known that

$$Tr(\alpha^2) = [Tr(\alpha)]^2 = Tr(\alpha) \in GF(2),$$

$$Tr(\alpha + \beta) = Tr(\alpha) + Tr(\beta),$$

$$Tr(1) = 1.$$

For each $\gamma \in GF(2^{2m-1})$, we define an alternating bilinear form B_γ on V(2m, 2) by

$$B_\gamma(\alpha, \beta) = Tr(\alpha\beta\gamma^2) + Tr(\alpha\gamma)\cdot Tr(\beta\gamma),$$

$B_\gamma(\alpha, \xi) = \mathrm{Tr}(\alpha\gamma)$,

$B_\gamma(\xi, \xi) = 0$,

for all $\alpha, \beta \in GF(2^{2m-1})$ and $\xi \in GF(2)$, extending linearly to $V(2m, 2)$. We shall show that

$$K = \{B_\gamma \mid \gamma \in GF(2^{2m-1})\}$$

is a Kerdock set. It obviously has the right number of elements, so it only needs to be shown that $B_\gamma + B_\delta$ is non-singular when $\gamma \neq \delta$. Suppose then that $B_\gamma + B_\delta$ is singular. Since the kernel of $B_\gamma + B_\delta$ must have even dimension, there exists $\alpha \in GF(2^{2m-1})\backslash\{0\}$ which is also in the kernel; it follows that

$$\mathrm{Tr}(\alpha\gamma) + \mathrm{Tr}(\alpha\delta) = 0$$

and $\mathrm{Tr}(\alpha\beta\gamma^2) + \mathrm{Tr}(\alpha\beta\delta^2) + \mathrm{Tr}(\alpha\gamma)\cdot\mathrm{Tr}(\beta\gamma) + \mathrm{Tr}(\alpha\delta)\cdot\mathrm{Tr}(\beta\delta) = 0$ for all $\beta \in GF(2^{2m-1})$. If $\mathrm{Tr}(\alpha\gamma) = \mathrm{Tr}(\alpha\delta) = 0$, then $\mathrm{Tr}(\alpha\beta(\gamma + \delta)^2) = 0$ for all $\beta \in GF(2^{2m-1})$ and we conclude that $\gamma = \delta$. On the other hand, if $\mathrm{Tr}(\alpha\gamma) = \mathrm{Tr}(\alpha\delta) = 1$, then $\mathrm{Tr}\{\beta(\gamma + \delta)(\alpha[\gamma + \delta] + 1)\} = 0$ for all $\beta \in GF(2^{2m-1})$ and we again deduce that $\gamma = \delta$. This establishes that K is a Kerdock set.

4.3. Kerdock codes

Suppose that B is an alternating bilinear form on $V(2m, 2)$ and that Q and Q' are two quadratic forms associated with B as in §4.1; it is then an easy exercise to show that Q + Q' is one of the 2^{2m} linear forms on $V(2m, 2)$. Thus, there are 2^{2m} quadratic forms associated with each alternating bilinear form on V(2m, 2). The count of alternating bilinear forms in §4.2 shows that there are exactly $2^{m(2m+1)} = 2^{2m} \cdot 2^{m(2m-1)}$ elements in the set $\mathcal{Q}$ of all quadratic forms on V(2m, 2).

We may represent each $Q \in \mathcal{Q}$ as a vector in $V(2^{2m}, 2)$ whose 2^{2m} coordinates 0 and 1 are the values $Q(\underline{x})$ over all $\underline{x} \in V(2m, 2)$. From these binary "words" of length 2^{2m}, we form a Kerdock code by fixing a Kerdock set $\mathcal{K}$ and taking as code words

$$\{Q,\ 1 + Q \mid \text{the bilinear form } B(Q) \in \mathcal{K}\}.$$

Since there are 2^{2m-1} bilinear forms in $\mathcal{K}$, 2^{2m} quadratic forms for each bilinear form, and 2 code words for each quadratic form, we have a total of 2^{4m} code words. Since the smallest number of 1's in a combination is $2^{2m-1} - 2^{m-1}$, the Kerdock code is found to be a binary $(2^{4m},\ 2^{2m},\ 2^{2m-1} - 2^{m-1})$-code.

5. EQUIANGULAR LINES IN R^d

5.1. Background

If a collection of n lines through a point O in R^d has the property that each pair of distinct lines determines the same angle ψ, the collection is called a set of equiangular lines. A unit vector p_i may be chosen along each line of such a collection in either of two directions; it follows that the inner product between distinct unit vectors is given by

$\langle p_i, p_j \rangle = \pm\cos\psi$.

The matrix of inner products may thus be rewritten as

$$P = \langle p_i, p_j \rangle = \begin{vmatrix} 1 & & & \pm\cos\psi \\ & \cdot & & \\ & & \cdot & \\ & & & \cdot \\ \pm\cos\psi & & & 1 \end{vmatrix} = I + C\cos\psi$$

where C is a symmetric $n \times n$ matrix with 0 along the main diagonal and ± 1 elsewhere. In order that the n lines span R^d, we may assume that $n > d$. Since there are now more directions than dimensions, the smallest eigenvalue of P is 0, and its multiplicity is at least $n - d$. Because C is obtained from P in the above manner, we find that the smallest eigenvalue of C is $-1/\cos\psi$, and its multiplicity is at least $n - d$. Conversely, suppose that C_n is a symmetric $n \times n$ matrix with 0 along the main diagonal and ± 1 elsewhere, that γ_0 is the smallest eigenvalue of C_n, and that γ_0 has multiplicity $n - d$. It follows that $C_n - \gamma_0 I$ is a symmetric positive-semidefinite $n \times n$ matrix of rank d, and thus that there is an $n \times d$ matrix U such that

(5.A) $C_n - \gamma_0 I = U \cdot U^T$.

Then n rows of U may be considered vertices in R^d, and (5.A) shows that $C_n - \gamma_0 I$ is the matrix of inner products for these vectors. Since the non-diagonal elements of $C_n - \gamma_0 I$ are all ± 1, we see that the n lines spanned by these vectors are equiangular.

Therefore, to find large sets of equiangular lines,

we look for large matrices C_n whose smallest eigenvalue has a large multiplicity. For instance, there is a 16 × 16 symmetric Hadamard matrix H_{16} which may be written as $H_{16} = C_{16} - I$ where $(C_{16} - I)^2 = 16I$; it follows that $(C_{16} - 5I)(C_{16} + 3I) = 0$, so the eigenvalues of C_{16} are 5 and -3 (with multiplicities 6 and 10 respectively). Proceeding in the manner described above, we discover a set of 16 equiangular lines in R^{16-10} $(=R^6)$.

5.2. Graphs and equiangular lines

The matrix C associated above with a set of equiangular lines may be interpreted as the adjacency matrix A of a graph via the transition

$$C = J - I - 2A,$$

i.e., 1 and -1 in C correspond to 0 and 1 in A respectively. For instance, the adjacency matrix A_{10} of the Petersen graph was found in §3.2 to satisfy $A_{10}^2 + A_{10} - 2I = J$, $Aj = 3j$ so the corresponding C_{10} satisfies $C_{10}^2 = 9I$. The multiplicities of the eigenvalues 3 and -3 of C_{10} are both 5. Using the procedure of §5.1, we discover a set of 10 equiangular lines in R^{10-5} $(=R^5)$.

The correspondence between graphs and sets of equiangular lines is not 1-1, due to the fact that we originally could have chosen the unit vectors along the lines in either of two ways. Examining how a choice of $-p_i$ instead of p_i is reflected in the adjacency matrix C, we define a "switching relation" on (±1)-adjacency matrices: C and C' are related if there is a diagonal matrix D whose diagonal elements are ±1 such that C' = DCD. This switching relation is easily seen to be an equivalence relation. The switching classes of graphs are now found to be in 1-1 correspondence with the sets of equiangular lines. For graphs with 3 vertices, there are two switching classes, consisting of graphs with an even number of edges or graphs with an odd number of edges (the latter are called odd 3-graphs). Graphs with 4 vertices are distributed among switching classes according as the number of odd 3-subgraphs is 0, 2, or 4.

5.3. Two-graphs

If $\Omega^{(i)}$ denotes the collection of i-subsets of the set Ω, a graph is occasionally defined as a pair (Ω, f) where f: $\Omega^{(2)} \to \{1, -1\}$ is an adjacency function. Similarly, we define a <u>two-graph</u> to be a pair (Ω, g) where g: $\Omega^{(3)} \to \{1, -1\}$ is a map satisfying

$$g(\alpha, \beta, \gamma)\cdot g(\alpha, \beta, \delta)\cdot g(\alpha, \gamma, \delta)\cdot g(\beta, \gamma, \delta) = 1$$

for all distinct $\alpha, \beta, \gamma, \delta \in \Omega$. If a triple $\{\alpha, \beta, \gamma\}$ is called <u>coherent</u> when $g(\alpha, \beta, \gamma) = -1$, the above condition (essentially the cohomology condition $\delta g = 1$) allows that, whenever we know which triples containing a particular vertex α are coherent, we know the coherent triples for the whole two-graph.

Given a graph (Ω, f), we may define a two-graph (Ω, g)

by $g(\alpha, \beta, \gamma) := f(\alpha, \beta)\cdot f(\beta, \gamma)\cdot f(\gamma, \alpha)$; the coherent triples of the two-graph are precisely the odd 3-subgraphs of the original graph. On the other hand, if we start with a two-graph (Ω, g), we may associate with it the switching class

$$\{(\Omega, f) \mid f(\alpha, \beta)\cdot f(\beta, \gamma)\cdot f(\gamma, \alpha) = g(\alpha, \beta, \gamma)\}$$

of ordinary graphs. In this way, we discover

THEOREM 5.1. *There is a 1-1 correspondence between*

(i) *two-graphs*,

(ii) *switching classes of graphs, and*

(iii) *sets of equiangular lines.*

COROLLARY. *If a graph* (Ω, f) *belongs to the switching class of the two-graph* (Ω, g), *then* $\mathrm{Aut}(\Omega, f)$ *is a subgroup of* $\mathrm{Aut}(\Omega, g)$.

5.4. 2-transitive two-graphs

Shult employs this last observation to construct a two-graph with a 2-transitive automorphism group. Suppose that (Ω, f_0) is a graph, $x \in \Omega$, Γ_x = {vertices adjacent to x} and Δ_x = {vertices non-adjacent to x}. We suppose furthermore that $\mathrm{Aut}(\Omega, f)$ is transitive on Ω, and that there exist automorphisms h_1 and h_2 of the subgraphs Γ_x and Δ_x respectively such that $h_1(y)$ is adjacent to $h_2(z)$ precisely when y and z are non-adjacent. Under these conditions, it follows that there is a two-graph $(\Omega \cup \{\omega\}, g)$ with a 2-transitive automorphism group. In fact, the switching class of the graph $(\Omega \cup \{\omega\}, f)$ where

$$f(y, z) = \begin{cases} 1 & \text{if either } y = \omega \text{ or } z = \omega, \\ f_0(y, z) & \text{otherwise} \end{cases}$$

may be shown to contain an element interchanging ω and x; since the original assumptions make it clear that the stabilizer of ω is transitive on the remaining vertices, we see that the switching class of $(\Omega \cup \{\omega\}, f)$ has an automorphism group which is 2-transitive on $\Omega \cup \{\omega\}$. Theorem 5.1 now establishes the existence of the required two-graph.

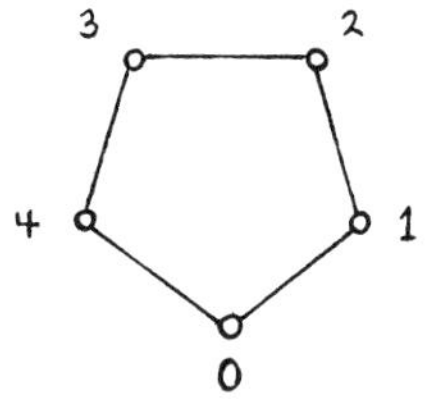

FIGURE A

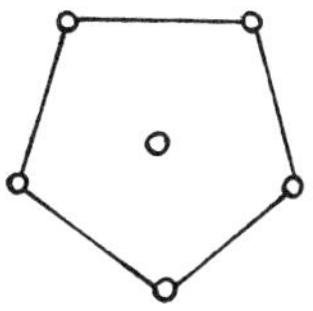
FIGURE B

For example, the pentagon in figure A has a transitive automorphism group. If h_1 interchanges the two vertices of $\Gamma_0 = \{1, 4\}$ and h_2 is chosen to be the identity on $\Delta_0 = \{2, 3\}$, then we find that our conditions are fulfilled. It follows that there is a two-graph with six vertices whose automorphism group is 2-transitive; this two-graph is associated with the switching class of the graph representing the diameters of the icosahedron (figure B), and its automorphism group is A_5. This procedure generalizes to produce a two-graph with $q + 1$ vertices and an automorphism group $PSL(2, q)$ which is 2-transitive.

Another example may be obtained from the symplectic graph on $V(2m, 2)\backslash\{\underline{o}\}$ introduced in §4.1. We saw that $Sp(2m, 2)$ was transitive on the vertices of that graph. If $\underline{x}$ is a vertex, then $\Gamma_{\underline{x}} = \{\underline{y} \neq \underline{0} \mid B(\underline{x}, \underline{y}) = 0\}$ and $\Delta_{\underline{x}} = \{\underline{z} \mid B(\underline{x}, \underline{z}) = 1\}$. If h_1 is the mapping

$$h_1 : \Gamma_{\underline{x}} \to \Gamma_{\underline{x}}; \quad \underline{y} \mapsto \underline{x} + \underline{y}$$

and h_2 is chosen to be the identity on $\Delta_{\underline{x}}$, then our conditions are fulfilled and there exists a two-graph on 2^{2m} vertices with a 2-transitive automorphism group $Sp(2m, 2) \cdot V(2m, 2)$. If the vertices are the elements of $V(2m, 2)$, we find that $\underline{x}$, $\underline{y}$, $\underline{z}$ are coherent exactly when $B(\underline{x}, \underline{y}) + B(\underline{y}, \underline{z}) + B(\underline{z}, \underline{x}) = 0$.

As a final example, we consider the orthogonal graph on the quadric

$$\Omega^{\varepsilon} = \{\underline{x} \in V(2m, 2)\backslash\{\underline{o}\} \mid Q^{\varepsilon}(\underline{x}) = 0\}$$

where $\varepsilon = \pm 1$, as introduced in §4.1. The graph has $O^{\varepsilon}(2m, 2)$ as a transitive automorphism group, and a construction as above yields a two-graph with $2^{m-1}(2^m + \varepsilon)$ vertices and a 2-transitive automorphism group $Sp(2m, 2)$.

For further examples such as those involving Conway's group $\cdot 3$, we refer to [13].

6. REGULAR TWO-GRAPHS

6.1. Definitions and properties

We proceed with notions due to G. Higman [11], [16]. A two-graph (Ω, g) is called <u>regular</u> if each pair α, β of vertices is contained in exactly r coherent triples. In terms of the ideas of §5, it may be shown that this is equivalent to insisting that any matrix C (of size $n \times n$) in the switching class associated with (Ω, g) has exactly two eigenvalues, say $\rho_1 > \rho_2$, and $(C - \rho_1 I)\cdot(C - \rho_2 I) = 0$. We shall set forth several consequences of this latter definition. The multiplicities of the eigenvalues ρ_1 and ρ_2 will be denoted by μ_1 and μ_2 respectively, so

$$\text{(6.A)} \qquad \begin{aligned} \mu_1 + \mu_2 &= n = 1 - \rho_1\rho_2, \\ \mu_1\rho_1 + \mu_2\rho_2 &= 0. \end{aligned}$$

It follows that

(6.B) If $\rho_1 + \rho_2 \neq 0$, then ρ_1 and ρ_2 are odd integers.

(6.C) If $\rho_1 + \rho_2 = 0$, then $\rho_1 = -\rho_2 = \sqrt{n - 1}$ and $C^2 = (n - 1)I$; furthermore, it may be shown that $n \equiv 2 \pmod 4$ and that $n - 1$ is the sum of two squares. Examples have been constructed when $n - 1 = p^k \equiv 1 \pmod 4$, and also for $n = 226$; the first unknown possibility occurs when $n = 46$.

Solving (6.A) for μ_2, we find

$$\mu_2 = \rho_1^2 - \frac{\rho_1^3 - \rho_1}{\rho_1 - \rho_2}$$

and we deduce

(6.D) If $\rho_1 = 3$, then $n = 10, 16,$ or 28. If $\rho_1 = 5$, then $n = 16, 26, 36, 76, 96, 126, 176,$ or 276. Moreover, examples of each possibility are known, except for $\rho_1 = 5$ and $n = 76$ or 96.

Furthermore

(6.E) $2\rho_1 \leq 3\rho_2 - \rho_2^3$.

This is a consequence of the following (cf. also [21])

THEOREM 6.1. <u>The number of equiangular lines in</u> R^d <u>is at most</u> $\frac{1}{2}d(d+1)$.

PROOF. Suppose $p_1, \ldots, p_n$ are unit vectors along n equiangular lines in R^d, so $\langle p_i, p_j\rangle = \pm\alpha$ if $i \neq j$. Let

$$P_i : R^d \to R^d, \quad \underline{x} \mapsto \langle \underline{x}, p_i\rangle p_i$$

be the projection onto the vector p_i; then P_i is a symmetric linear map ($1 \leq i \leq n$). Moreover, if $i \neq j$,

$$\begin{aligned} Tr(P_iP_j) &= \sum_k \langle P_iP_j\underline{e}_k, \underline{e}_k\rangle \\ &= \sum_k \langle p_i, \underline{e}_k\rangle\cdot\langle p_i, p_j\rangle\cdot\langle p_j, \underline{e}_k\rangle \\ &= \langle p_i, p_j\rangle^2 = \alpha^2. \end{aligned}$$

Since an inner product of two symmetric linear maps Q and R is given by tr(QR), we may deduce from this that $P_1, \ldots, P_n$ are linearly independent. Since the space of symmetric linear maps on R^d has dimension $\frac{1}{2}d(d+1)$, it follows that $n \leq \frac{1}{2}d(d+1)$, as required.

6.2. Unitary two-graphs

Let q be an odd prime power, $PG(2, q^2)$ the projective plane obtained from $V(3, q^2)$, H a non-degenerate hermitian form on $V(3, q^2)$, and

$$\Omega := \{(\underline{x}) \in PG(2, q^2) \mid H(\underline{x}, \underline{x}) = 0\}$$

the quadric associated with H. It is known that $|\Omega| = q^3 + 1$ and that $P\Gamma U(3, q^2)$ acts 2-transitively on Ω. Suppose Δ is the set of distinct $(\underline{x})$, $(\underline{y})$, $(\underline{z})$ in Ω such that

$$H(\underline{x}, \underline{y}) \cdot H(\underline{y}, \underline{z}) \cdot H(\underline{z}, \underline{x}) = \begin{cases} \text{a square in } GF(q^2) \text{ if } \\ q \equiv -1 \pmod 4, \\ \\ \text{a non-square in } GF(q^2) \text{ if } \\ q \equiv 1 \pmod 4. \end{cases}$$

Then (Ω, Δ) is a regular two-graph, with

$$n = q^3 + 1, \quad \mu_1 = q^2 - q + 1, \quad \mu_2 = q(q^2 - q + 1),$$

$$(C - q^2 I) \cdot (C + qI) = 0$$

(it is a slightly complicated process to establish these results about the parameters). Theorem 5.1 now implies that there is a set of $q^3 + 1$ equiangular lines in R^{q^2-q+1} with $\cos \psi = 1/q$. Thus, the maximum number of equiangular lines in R^d is at least $d\sqrt{d}$.

6.3. Equiangular lines

We now tabulate the maximum number $n(d)$ of equiangular lines in R^d as follows:

d	n(d)	reference
2	3	
3-4	6	
5	10	5.2
6	16	5.1
7-14	28	5.4, 6.1
15	36	5.4
16	40	
17-18	48	

d	n(d)	reference
19	72	
20	90	
21	126	$U(3, 5^2)$
22	176	Higman-Sims
23-42	276	constructed below
43	344	$U(3, 7^2)$

6.4. The 276-two-graph

A non-trivial regular two-graph on 276 vertices may be constructed as follows [5]. From (6.A), (6.B), and (6.C), we deduce that the only possible eigenvalues ρ_1, ρ_2 of C are 55 and -5. Now the complementary Golay code consists of the 243 vectors, having Hamming distances 6 and 9, in a 5-dimensional subspace of V(11, 3). Let

$$PR_i: V(11, 3) \to GF(3), \quad (x_1, \ldots, x_{11}) \to x_i$$

be the projection onto the ith coordinate ($1 \le i \le 11$); there are 33 pairs (PR_i, x) where PR_i is one of these 11 projections and $x \in GF(3)$. Let the 276 vertices of a graph consist of these 33 pairs and the 243 elements of the complementary Golay code. Adjacency is defined by:

(PR_i, x) is adjacent to $(PR_j, y) \iff x = y$

(PR_i, x) is adjacent to $\underline{y} = (y_1, \ldots, y_{11})$

$\iff x = PR_i(\underline{y}) = y_i$

$\underline{x} = (x_1, \ldots, x_{11})$ is adjacent to $\underline{y} = (y_1, \ldots, y_{11})$

$\iff$ their Hamming distance is 9.

We find that the adjacency matrix C for this graph satisfies

$$(C + 5I)\cdot(C - 55I) = 0$$

so the two-graph associated with the switching class of

this graph is regular. This two-graph is the only non-trivial regular two-graph on 276 vertices, since the above process may be reversed. Another way to construct 276 equi-angular lines in R^{23} is to consider in the Leech lattice in R^{24} the 276 rhombi having a fixed diagonal at the second smallest distance occurring in the lattice.

REFERENCES

1. F.C. Bussemaker and J.J. Seidel. Symmetric Hadamard matrices of order 36. T.H. Report 70-WSK-02, Tech. Univ. Eindhoven, Netherlands (1970).
2. E.R. Berlekamp, J.H. van Lint, and J.J. Seidel. A strongly regular graph derived from the perfect ternary Golay code. A Survey of Combinatorial Theory 1973 (J.N. Shrivastava, editor), 25-30. North Holland Pub. Co.
3. P.J. Cameron and J.J. Seidel. Quadratic forms over GF(2). Proc. Kon. Ned. Akad. Wet. Ser. A, 76 (1973): 1-8.
4. J.M. Goethals. Some combinatorial aspects of coding theory. A Survey of Combinatorial Theory 1973 (J.N. Shrivastava, editor), 189-208. North Holland Pub. Co.
5. J.M. Goethals and J.J. Seidel. The regular two-graph on 276 vertices. Discrete Math. 12 (1975): 143-158.
6. M. Hall. Combinatorial Theory. Waltham, Mass. (1967).
7. A.J. Hoffman and R.R. Singleton. On Moore graphs with diameters 2 and 3. IBM J. Research Develop. 4 (1960): 497-504.
8. P.W.H. Lemmens and J.J. Seidel. Equiangular lines. J. Algebra 24 (1973): 494-512.
9. J.J. Seidel. Strongly regular graphs. Progress in Combinatorics 1969 (W.T. Tutte, editor), 185-197. Academic Press.
10. —— On two-graphs and Shult's characterization of symplectic and orthogonal geometries over GF(2). T. H. Report 73-WSK-02, Tech. Univ. Eindhoven, Netherlands (1973).
11. —— A survey of two-graphs. Proc. Int. Coll. Teorie Combinatorie, Acc. Naz. Lincei, Rome (to appear).
12. —— Graphs and two-graphs. Fifth Southeastern Conf. on Combinatorics, Graph Theory & Computing. Utilitas Math. Publ. Inc., Winnipeg (1974): 125-143.
13. E.E. Shult. Characterizations of certain classes of graphs. J. Comb. Theory (B), 13 (1972): 1-26.

14. —— The graph-extension theorem. Proc. Amer. Math. Soc. 33 (1972): 278-284.
15. N.J. Sloane and J.J. Seidel. A new family of non-linear codes obtained from conference matrices. Int. Conf. on Combinatorial Math. 175 (1970): 363-365.
16. D.E. Taylor. Regular two-graphs. Proc. London Math. Soc. (to appear).
17. J.H. van Lint. Coding theory. Springer Lecture Notes 201.
18. —— Combinatorial theory seminar. Springer Lecture Notes 382.
19. J.H. van Lint and J.J. Seidel. Equilateral point sets in elliptic geometry. Proc. Kon. Ned. Akad. Wet. Ser. A, 69 (1966): 335-348.
20. J.S. Wallis. Hadamard matrices. Springer Lecture Notes 292.
21. Ph. Delsarte, J.M. Goethals, J.J. Seidel. Bounds for systems of lines, and Jacobi polynomials. Philips Res. Repts. 30 (1975): 91* - 105*.

DISTRIBUTIVE QUASIGROUPS

K. Strambach

We consider the arithmetic mean, which is the following mapping of the real line:

$$(x, y) \to \frac{x + y}{2} = x \circ y.$$

We first collect some significant properties of the arithmetic mean. Clearly, the following are valid:

(0) $\circ$ is idempotent.

(1) $(a\circ b)\circ c = (a\circ c)\circ(b\circ c)$.

(2) $a\circ(b\circ c) = (a\circ b)\circ(a\circ c)$.

(3) The equation $a \circ \xi = b$ has a unique solution for given a and b.

(4) The equation $\xi \circ a = b$ has a unique solution for given a and b.

(5) $a\circ b = b\circ a$ for all a and b.

We now reverse the situation and ask: to what extent does each single property determine the arithmetic mean?

Let D be a set with a multiplication $\circ$. We define mappings

$$a^D = (x \mapsto x \circ a) : D \to D.$$

Such a mapping will be called a right translation by a. If we assume property (1), then the right translations are endomorphisms. If the multiplication satisfies (1) and (3) then $(a \circ b)^D = (b^D)^{-1} a^D (b^D)$, and we can regard the set D as a class of conjugate elements in the group Σ generated by all right translations of D, where the multiplication $\circ$ can be regarded as conjugation in Σ. Thus the classification of structures with properties (1) and (3) can be reduced to a classification of groups Σ (generated by the right translations) and their classes of conjugate elements.

A set D with multiplication $\circ$ satisfying properties (1) to (5) will be called a commutative two-sided distributive quasigroup. The classification of such structures can be obtained rather easily. Here, every group Σ is the semidirect product B of a suitable abelian group A by a cyclic group $\langle\alpha\rangle$, such that the normalizer of $\langle\alpha\rangle$ in B is $\langle\alpha\rangle$ and A satisfies certain further restrictions. If, for instance, A is torsion-free, then A is 2-divisible and $\alpha^{-1}x\alpha = x^{\frac{1}{2}}$ for all $x \in A$. If A is a torsion group, then we get restrictions on the orders of the elements of A. As elements of the quasigroup D we then take the conjugacy classes of $\langle\alpha\rangle$ in B; as multiplication, we take conjugation. In the classical case, we have $\Sigma = \mathbb{R}\langle\alpha\rangle$ with $\alpha^{-1}x\alpha = \frac{x}{2}$; the set S of right translations is

$$S = \{x \to \frac{x}{2} + b;\ b \in \mathbb{R}\}.$$

The mapping $\{x \to \frac{x}{2} + b\} \mapsto b: S \to \mathbb{R}$ yields an isomorphism between the "conjugation" multiplication and the multiplication $(x, y) \to \frac{x + y}{2}$.

We can obtain generalized means on the real line, which are mappings of the form

$$(x, y) \mapsto px + (1 - p)y: \mathbb{R} \to \mathbb{R} \text{ with } 0 \neq p \neq 1.$$

These behave just as nicely (up to commutativity) as the ordinary arithmetic mean. For this reason we shall henceforth only regard algebraic structures here which satisfy at best conditions (1) to (4).

A set D carrying an algebraic structure which satisfies (1) to (4) will be called a <u>two-sided quasigroup</u>. A classification of two-sided quasigroups is in general very complicated and still an open question. The best result about two-sided quasigroups, due to Bernd Fischer [3], says the following: if D is a finite two-sided quasigroup, and Σ is the group generated by all right translations, then Σ is a solvable (finite) group. It seems, apart from the finiteness, that this theorem depends essentially on the validity of both distributive laws.

The right-distributive law is equivalent to the following statement about the set of right translations D^D of D: if α is an element of D^D, then D^D is a system of representatives for the right cosets of the centralizer $Cs(\alpha)$ of a right translation α, so that Σ is decomposable into cosets $Cs(\alpha)d_i$ with $d_i \in D^D$.

The left distributive law is equivalent to the following relation between any three right translations α, β, γ:

(*) $(\beta^{-1}\gamma)\alpha(\gamma^{-1}\beta) = (\alpha^{-1}\gamma\beta^{-1})\alpha(\beta\gamma^{-1}\alpha)$.

Now we shall show that there exist simple groups which have a conjugate class of elements such that this class, together with conjugation as multiplication, satisfies conditions (1), (3), and (4), but not (2).

Let Σ be the group $PSL_2(\mathbb{R})$. Then Σ can be interpreted as the group of proper motions of the classical hyperbolic plane $\mathbb{H}$. Let L be a conjugate class of compact elements of Σ - that is, a conjugate class of hyperbolic rotations. In the set L, we define a multiplication $\circ$ by conjugation: $\alpha \circ \beta = \beta^{-1}\alpha\beta$ for all $\alpha \in L$. We claim that L, together with the multiplication $\circ$, fulfils conditions (1), (3), and (4); we shall call such a structure a <u>right-distributive quasigroup</u>.

Condition (1) is a consequence of the fact that L form a complete class of conjugate elements. Similarly conditi (4). To solve the equation $\xi \circ \alpha = \beta$ for given α, β, we compute $\xi = \alpha\beta\alpha^{-1}$. The only serious question we are left with is whether L satisfies condition (3). Every element (rotation) of L is uniquely determined (mod π) by its centre, its angle, and its sense of rotation. To prove that the equation $\gamma \circ \xi = \delta = \xi^{-1}\gamma\xi = \delta$ has a unique solution for a given γ and δ we must show that L contains a unique rotation which maps the centre f_γ of γ onto the centre f_δ

of δ. To do this, we erect the perpendicular bisector S of the segment $f_\gamma f_\delta$. The line joining f_γ and f_δ separates the plane into two half-planes. Hyperbolic rotations λ with centres on S, angles $\leq\pi$, that map f_γ onto f_δ will have the same sense of rotation exactly if their centres lie in the same (closed) half-plane. We now let H be the closed half-plane in which these rotations have the same rotation sense as the elements of L. To every point s of $S \cap H$, we can now associate a unique rotation $\phi(s)$ such that $f_\gamma^{\phi(s)} = f_\delta$. We can also associate with s the rotation angle $\omega(s)$ of $\phi(s)$. As s goes from $0 = S \cap [f_\gamma f_\delta]$ to infinity, $\omega(s)$ is a continuous strictly decreasing function with $\omega(0) = \pi$ and $\lim_{s\to\infty} \omega(s) = 0$. Hence, there exists a unique point s_0 on $S \cap H$ such that $\omega(s_0)$ is equal to the angle of the elements belonging to L. Then $\xi = \phi(s_0)$ is the unique element of L which solves the equation $\gamma \circ \xi = \delta$. Hence L is a right-distributive quasigroup which has a simple group as the group generated by all right translations. Moreover, L is a topological quasigroup - that is, both the multiplication and the translations are continuous. The topological space of L is the plane $\mathbb{R}^2$, and the quasigroup L is realized on a manifold. (If L were left-distributive, then computation would show that condition (*) is violated.)

We observe that two quasigroups L_1 and L_2 consisting of full classes of hyperbolic motions are isomorphic exactly if there exists some element σ in the full group PGL($\mathbb{R}$)

of hyperbolic motions for which $L_1^\sigma = L_2$. (If we consider two rotations $\alpha_i \in L_i$ with the same centre, then L_1 is isomorphic to L_2 if and only if $\alpha_2 = \alpha_1^{\pm 1}$.)

Every quasigroup L has the property that any two distinct right translations generate an automorphism group which contains all right translations, and so the whole group Σ. We shall call quasigroups with this property minimal and classify them in the case that both the quasigroup and the group Σ generated by right translations are locally compact.

In our next examples, we show that also non-solvable compact Lie groups can occur as groups generated by the right translations of a right-distributive quasigroup.

Let Δ be one of the following groups: $PSO_8(\mathbb{R})$, $Spin_8(\mathbb{R})$, $PSO_8(\mathbb{C})$, $Spin_8(\mathbb{C})$, and $PSO_8(\mathbb{R}, h)$, where h is a quadratic form of index 4 and τ is a triality automorphism of order 3. (The outer automorphisms of simple Lie groups can be obtained as automorphisms of the Dynkin diagrams associated with any simple Lie algebra. All the groups listed above have the same complex Lie algebra and the same Dynkin graph:

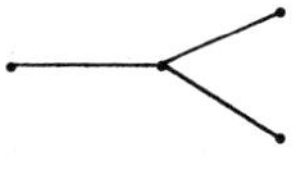

(An automorphism of Δ is called a triality automorphism if it has order 3 and permutes the edges of the Dynkin graph cyclically.)

We take Σ to be the semidirect product of Δ with the group $<\tau>$. If we take for L the set of elements formed by the conjugate class of $<\tau>$ in Σ, and prescribe multiplication by conjugation, we obtain a topological right-distributive quasigroup which is not left-distributive.

To prove this, we need only show that the equation $\alpha \circ \beta = \beta$ has a unique solution for given α and $\beta \in L$. This is equivalent to the fact that every coset of a centralizer $Cs(\alpha)$ of some $\alpha \in L$ contains exactly one right translation out of Σ. We shall show this for the case that Δ is isomorphic to one of $PSO_8(\mathbb{R})$, $PSO_8(\mathbb{C})$, or $PSO_8(\mathbb{R}, h)$. For in these cases we may regard Σ as a collineation group of an octonian plane over the classical octonians or over certain split octonian algebras. In the first case, Σ operates on the classical octonian plane; in the two other cases, Σ operates on octonian Moufang-Hjelmslev planes (as shown by Springer and Veldkamp [9]).

In every case, Δ leaves three points e_0, e_1, e_3 of a coordinate quadrangle $V = \{e_0, e_1, e_2, e_3\}$ fixed; we can choose the fourth point e_2 so that τ operates on V in such a way that $e_0^\tau = e_0$, $e_1^\tau = e_2$, $e_2^\tau = e_3$, and $e_3^\tau = e_1$. All elements of a coset of the centralizer $Cs(\tau)$ of the element τ map e_1 onto the same point of the orbit e_2^Δ; distinct cosets of $Cs(\tau)$ map e_1 onto distinct points of e_2^Δ, since the centralizer of τ in Δ leaves the quadrangle V pointwise fixed and is isomorphic to an exceptional Lie group of type G_2. The elements of L have the form $\lambda^{-1}\tau\lambda$ with $\lambda \in \Delta$. Because $e_1^{\lambda^{-1}\tau\lambda} = e_2^\lambda$, the element $\lambda^{-1}\tau\lambda$ lies in the coset

$Cs(\tau)\lambda$. If the elements $\rho^{-1}\tau\rho$ and $\lambda^{-1}\tau\lambda$ are contained in the same right coset, then we have

$$(\rho^{-1}\tau\rho)(\lambda^{-1}\tau^{-1}\lambda) = \rho^{-1}\tau(\rho\lambda^{-1})\tau^{-1}\lambda = \rho^{-1}\rho\lambda^{-1}\lambda = 1$$

because $\rho\lambda^{-1}$ belongs to $Cs(\tau)$. Hence every coset of $Cs(\tau)$ contains exactly one element of L, and we have shown that L is a right-distributive quasigroup. I can prove that L violates the left-distributive law by using Freudenthal's arguments from "Oktaven, Oktaven-geometrie und Liegruppen" [4]. We omit the proof here.

Since all triality automorphisms of Δ are conjugate, we obtain only one quasigroup for each of the above-mentioned four groups.

The right-distributive quasigroups given are realized on manifolds and are examples of analytical quasigroups; their associated groups generated by right translations are Lie groups. We are left with the following question: how typical are these examples within a classification of quasigroups on manifolds? In fact, it turns out later that the examples are very essential.

For Hopf's investigations tell us that every non-degenerate continuous multiplication on a manifold has rather strong consequences for the algebraic invariants of that manifold. For instance, a topological compact orientable manifold carrying a topological quasigroup is a Hopf manifold in the sense of Borel. From Borel we know that the cohomology ring whose coefficients lie in a field K of characteristic zero is an exterior algebra of a vector space over K. In addition, if D is a topological m-dimensional manifold, then the Poincaré duality enables us to show that the nth Betti number β_n is equal to $\binom{m}{n}$.

We now turn our attention to a type of multiplication on topological spaces which includes the quasigroup multiplications already discussed. We shall show that this type of multiplication leads to very strong restrictions on the homotopy groups of the manifolds admitted by the multiplication.

DEFINITION. If D is a topological space, then we shall call a continuous multiplication $\mu: D \times D \to D$ idempotent if $\mu(x, x) = x$ for all points $x \in D$. The multiplication μ is called regular idempotent if it is idempotent and if every translation $x \mapsto \mu(x, a)$ and $x \mapsto \mu(a, x)$ is a homeomorphism of D. It is clear that every quasigroup multiplication is regular idempotent. We now lead up to a theorem about regular multiplication by again studying a number of examples.

We know that $\mathbb{R}^n$ gives rise to a regular multiplication defined as follows: If x, y are vectors in $\mathbb{R}^n$, then $x \circ y = px + (1 - p)y$, with $0 \neq p \neq 1$ and $p \in \mathbb{R}$. The multiplication of x by the real number p can be regarded as on automorphism p acting on $\mathbb{R}^n$; clearly, $1 - p$ is then also an automorphism on $\mathbb{R}^n$.

We can generalize this situation to other topological abelian groups to obtain idempotent multiplications on a number of topological spaces. We shall also show later that it appears necessary for regular idempotent multiplications to exist that we find some automorphism α with the property that $1 - \alpha$ is also an automorphism.

Suppose we have an n-dimensional torus group T^n. Then the group of continuous automorphisms of T^n is the discrete group $U_n(\mathbb{Z})$ consisting of $(n \times n)$-matrices with determinant ± 1 and integers as elements. We can define a two-sided right-distributive quasigroup on T^n if we can find an automorphism α in $U_n(\mathbb{Z})$ such that (identity-α) is also an automorphism. Thus we look for matrices of $U_n(\mathbb{Z})$ such that $\det(\iota - \alpha) = \pm 1$. In $U_n(\mathbb{Z})$ there exist infinitely many such matrices. For instance, in $U_3(\mathbb{Z})$, we can find the matrices

$$(*) \quad \alpha = \alpha(a) = \begin{pmatrix} a & -1 & 0 \\ 2(1 - a) & 2 & 1 \\ 1 & 0 & 1 \end{pmatrix} \quad \text{for any } a \in \mathbb{Z}.$$

For these matrices we have $\det(\alpha) = \det(i - \alpha) = 1$ and trace $(\alpha) = a + 3$. Thus different choices of a will yield matrices which are not conjugate, and thus we get infinitely many non-isomorphic two-sided quasigroups on T^3.

We can also use the matrices defined in (*) to obtain examples of $\beta \in V_{3n}(\mathbb{Z})$ $(n \geq 1)$ with $\det(i - \beta) = 1$ if we arrange matrices $\alpha(a_i)$ of the form given by (*) along the main diagonal and then complete the matrix by filling in zero's in the remaining entries. The trace of the matrix

β so constructed will be $\sum_{i=1}^{n} a_i + 3n$. Thus we see that on every torus there exist infinitely many quasigroup operations.

The same construction principle also shows the existence of several regular idempotent multiplications on compact topological spaces which are not manifolds. Let S be a one-dimensional a-adic solenoid - that is, a one-dimensional connected compact abelian group; the character group X of S is then a subgroup of the discrete additive group of rational numbers. We choose S in such a manner that both the integer n and the integer n - 1 can occur as denominators of elements of X. Since the continuous automorphism groups of S and X are isomorphic, the mappings $x \to \frac{x}{n}$ and $x \to (n - 1)x$ are automorphisms of S and the assignment

$$\mu_n(x, y) = \frac{x}{n} + (1 - \frac{1}{n})y: S \times S \to S$$

defines a quasigroup multiplication on S.

The following theorem shows that the construction principle used above is no accident, but a natural condition:

THEOREM. *Let* T *be a full subcategory of the category Top of topological spaces and continuous mappings, which is closed relative to the finite products and which contains the one-point space. Let* π: T $\to$ Gr *be a (covariant) functor from* T *to the category of groups (and group homomorphisms) such that* $\pi(x \times y)$ *is naturally isomorphic to* $\pi(x) \times \pi(y)$.

If $m: D \times D \to D$ *is an idempotent multiplication on* $D \in T$, *then there exists an endomorphism* f *of* $\pi(D)$ *such that the group morphism* $M: \pi(D) \times \pi(D) \to \pi(D)$ *induced by* m *is given by*

$$M(\alpha, \beta) \to f(\alpha) + (1 - f)(\beta).$$

(We write the group $\pi(D)$ additively without assuming commutativity.) *If, in addition,* m *is a regular idempotent multiplication on* D, *then both* f *and* $(1 - f)$ *are group automorphisms and* $\pi(D)$ *is an abelian group. Finally, if* $\pi(D)$ *is cyclic, then it can neither be infinite nor of even order.*

In particular, let I be the subcategory of arcwise-connected topological spaces. We let π be any homotopy functor π_n. Then the theorem tells us that any fundamental group of a space admitting a regular idempotent multiplication is abelian and, in particular, that no sphere can carry such a multiplication.

For π, we can also take other homotopy functors. An instance of this is the so-called torus homotopy groups which in general are non-abelian even for $r = 1$. However, if a topological space admits a regular idempotent multiplication, then these groups are also abelian. Since the so-called Whitehead products correspond to the commutators of the torus homotopy groups, all the Whitehead products between the elements of homotopy groups of D must always be zero and the space D admitting μ is n-simple for any $n \geq 1$.

We might also let π be a more exotic functor, such as the functor $X \to$ Group Hom (SZ, X), where Group Hom (SZ, Z) is the group of homotopy classes of continuous mappings from a reduced (Freudenthal) suspension of a fixed (arcwise-connected) space Z to X.

We can state

COROLLARY 1. Let D be a two-dimensional topological manifold whose fundamental group is finitely generated. (This is of particular interest if the manifold is compact.) The surface D admits a regular idempotent multiplication exactly if D is homeomorphic to either the plane $\mathbb{R}^2$ or the torus $S^1 \times S^1$.

This corollary is a consequence of the fact that only 0, $\mathbb{Z}$, $\mathbb{Z}/2\mathbb{Z}$, or $\mathbb{Z} \times \mathbb{Z}$ occurs as an abelian fundamental group.

From Borel [2], we are acquainted with the homology groups of compact manifolds admitting regular idempotent multiplications, with coefficients taken in a field of characteristic 0. If D is a three-dimensional compact orientable manifold admitting one of our multiplications, we can compute the homology groups with coefficients in the ring $\mathbb{Z}$ of integers.

We can show the following:

COROLLARY 2. Let D be a connected compact orientable manifold admitting a regular idempotent multiplication. Then

$H_0(D; \mathbb{Z}) = H_3(D; \mathbb{Z}) = \mathbb{Z}$,

$H_1(D; \mathbb{Z}) = H_2(D; \mathbb{Z}) = \mathbb{Z} \times \mathbb{Z} \times \mathbb{Z}$.

This corollary follows from the Poincaré-duality and from a theorem due to Reidemeister (compare Hopf [6]) by which every abelian fundamental group of a compact three-dimensional manifold is either cyclic or isomorphic to $\mathbb{Z} \times \mathbb{Z} \times \mathbb{Z}$. It seems that the three-dimensional torus is the only compact three-dimensional manifold which admits a regular idempotent multiplication.

We have no complete survey about the homology ring of compact four-dimensional manifolds admitting regular idempotent multiplications with coefficients in $\mathbb{Z}$. However, let us make the following

DEFINITION. A compact manifold M is called 2-sphere-torsion free if the homology class of any continuous image of the 2-sphere in M is either 0 or of infinite order.

Now we can state:

COROLLARY 3. Let D be a 4-dimensional connected compact orientable manifold, which admits a regular idempotent multiplication and which is 2-sphere-torsion free. Then the following are true:

$H_0(D; \mathbb{Z}) = H_4(D; \mathbb{Z}) = \mathbb{Z}$,

$H_1(D; \mathbb{Z}) = H_3(D, \mathbb{Z}) = \mathbb{Z} \times \mathbb{Z} \times \mathbb{Z} \times \mathbb{Z}$,

$H_2(D: \mathbb{Z}) = \mathbb{Z} \times \mathbb{Z} \times \mathbb{Z} \times \mathbb{Z} \times \mathbb{Z} \times \mathbb{Z}.$

We now return to the theory of right-distributive quasigroups. In particular, we shall study topological quasigroups D, which are locally compact and locally connected. For these quasigroups induce the structure of a topological group in every group of automorphisms which has been provided with the compact open topology. From now on, we shall always take Σ to be the topological closure of that group which is algebraically generated by the right translations.

First, we state a rather general lemma.

LEMMA. Let G be a locally compact space, and let $(x, y) \to x \circ y : G \times G \to G$ be a continuous multiplication, such that all translations are homeomorphisms. Let CO(G, G) be the space of all continuous mappings from G to G with the compact open topology. Let $\mathcal{G}$ be the subset of CO(G, G) consisting of all right translations $\rho_a: x \to x \circ a$ with $a \in G$. Then

(1) $\mathcal{G}$ is closed in the space CO(G, G).

(2) The function $g \to \rho_g: G \to CO(G, G)$ is a homeomorphism of G into $\mathcal{G}$.

Half of this lemma yields some useful observations about the job of classifying quasigroups. Let D be a locally compact, locally connected, right-distributive quasigroup, and let Σ be the automorphism group topologically generated by all right translations. Then we have

(a) The right translations D^D of D form a conjugate class of Σ which is homeomorphic to D.

(b) For $\alpha \in D^D$, let $Cs(\alpha)$ be the centralizer of α in Σ. Then $Cs(\alpha) = \Sigma_a$ (stabilizer of Σ on a), and the space $\Sigma/Cs(\alpha) \approx D$.

(c) Σ has a trivial centre and operates transitively on D.

(d) The space D^D forms a slice of the space $\Sigma/Cs(\alpha)$ in Σ.

Our main aim is a classification of right-distributive topological quasigroups whose space is a manifold. We succeed with this classification if we either assume or are able to prove that the group Σ generated by right translations is "nice" - that is, if we know that Σ is a Lie group.

If, for instance, D is a topological quasigroup realized on a 1-dimensional manifold, then we can indeed prove that the group Σ is a Lie group.

From H. Kneser [7] we know that there are four types of 1-dimensional topological manifolds: the real line, the circle, the Alexandrov half-line, and the Alexandrov line. For a right-distributive quasigroup which is realized on either the real line or the Alexandrov line, the right translations are given by strictly monotonic surjective functions. However, from the work of M. Kneser and H. Kneser [8] we know that any two surjective, strictly increasing or strictly decreasing functions on an Alexandrov line, or on a half-line, have infinitely many points where these functions coincide. This yields a contradiction to the fact that the equations $a \circ x = b$ are uniquely solvable.

On the 1-sphere, no right-distributive quasigroup can exist due to the fact that $\pi_1(S_1) \simeq \mathbb{Z}$.

Suppose now we have $D \approx \mathbb{R}$. Then $\Sigma/\Sigma_a \approx \mathbb{R}$. Furthermore, if we analyse the behaviour of monotonic functions, we come up with the following:

PROPOSITION. Let D be a right-distributive quasigroup which is homeomorphic to $\mathbb{R}$. Then D contains no proper closed sub-quasigroups which contain more than one point.

It follows from this proposition that $\Sigma_{a,b} = 1$ if $a \neq b$. Also, Σ_a has a subgroup ψ of index at most 2 which operates freely on each of the two half-lines of $D \backslash a$. Hence, ψ admits an archimedian ordering, and is therefore an abelian subgroup of $\mathbb{R}$ which is closed in $\mathbb{R}$. Roughly speaking, it is for these reasons that Σ is the group

$$\{x \to r^u x + b \mid r \neq 0,\ u \in \mathbb{Z} \text{ and } b \in \mathbb{R}\}$$

and the right translations are the set of mappings

$$\{x \to rx + b \mid b \in \mathbb{R}\}.$$

From this, the following classification theorem follows very easily:

THEOREM. _Every right-distributive topological quasigroup D which is realized on a 1-dimensional connected manifold has the following form:_ $D \approx \mathbb{R}$ _and the multiplication_ $\circ$ _is given by_

$$x \circ y = px + (1 - p)y \text{ for some real number } p \text{ with } 0 \neq p \neq 1.$$

Two quasigroups are isomorphic (both topologically and abstractly) if and only if $p_1 = p_2$.

This theorem answers a question posed by Aczel [1]: is every right-distributive quasigroup on the real line two-sided distributive?

We now observe that all these 1 distributive quasigroups, with the exception of $\Sigma = \mathbb{R}\langle\alpha\rangle$ with $\alpha^{-1}x\alpha = -x$, have the property that any two distinct right translations topologically generate the whole group Σ generated by all right translations. We call right-distributive quasigroups with this property _minimal_. We ask whether we can classify all the minimal right-distributive quasigroups.

We can give such a classification under the additional assumptions that these quasigroups are locally compact, connected, locally connected and that the group Σ is a locally compact group. Under these conditions we can show that Σ is a Lie group. This is roughly due to the fact that every disconnected subgroup of a connected, locally compact group must lie in the centre of that group. The classification theorem about these minimal quasigroups may be formulated as follows:

THEOREM. Let D be a locally compact, connected, and locally connected minimal right-distributive quasigroup. Then the following possibilities exist for D:

(1) D is homeomorphic to $\mathbb{R}$ and the right translations are not involutions.

(2) Let $B \simeq \mathbb{R}^2 \cdot SO_2$ be the group of proper euclidean motions. Let D be a conjugate class of some euclidean rotation $\alpha \in SO_2$ in B such that the order of α is different from 2, 3, 4, or 6. Then, with respect to the multiplication

$$x \circ y = y^{-1}xy \quad \text{for} \quad x, y \in D$$

D forms a minimal two-sided distributive quasigroup which is homeomorphic to $\mathbb{R}^2$. Two quasigroups D_α and D_β defined in this way are isomorphic exactly if $\alpha = \beta$ or $\alpha = \beta^{-1}$. Also, $\Sigma \subset \mathbb{R}^2 \cdot \langle \bar{\alpha} \rangle$.

(3) Let B' be the usual extension of the euclidean translation group $\mathbb{R}^2$ by the euclidean group S consisting of spiral similarities - that is

$$S = \left\{ \begin{pmatrix} d^t \cos t & d^t \sin t \\ -d^t \sin t & d^t \cos t \end{pmatrix} \middle| \; d > 1 \text{ fixed and } t \in \mathbb{R} \right\}.$$

$B^1 = \mathbb{R}^2 \cdot S$ and $\mathbb{R}^2 \cap S = 1$. If D is the conjugate class of some element $\alpha \in S$ in B^1, then D forms a two-sided distributive quasigroup homeomorphic to $\mathbb{R}^2$ if we set

$x \circ y = y^{-1}xy$ for $x, y \in D$.

Two quasigroups D_α and D_β defined in this manner are isomorphic if and only if $\alpha = \beta^{\pm 1}$. Also, $\Sigma(D) \simeq \mathbb{R}^2 \cdot \langle\alpha\rangle$ and the connected component of Σ is isomorphic to the vector group $\mathbb{R}^2$.

(4) Let $\Delta = PSL_2(\mathbb{R})$. In Δ, let L_2 be the class of rotations which are conjugate to a given rotation α. Defining $x \circ y = y^{-1}xy$ for $x, y \in L_2$, we obtain a right-distributive, but not left-distributive quasigroup. Quasigroups L_{α_1} and L_{α_2} are isomorphic if and only if there exists some element δ in the group $PGL_2(\mathbb{R})$ such that $\delta^{-1}\alpha_1\delta = \alpha_2$.

Hence we see that the group $PSL_2(\mathbb{R})$ is the only locally compact group which leads to minimal, locally compact, connected, and locally connected quasigroups which are right, but not left, distributive.

In the remaining cases in which we have a classification of right-distributive connected quasigroups we must assume that the group Σ generated by right translations is a Lie group. Only in this case does our machinery work. For this purpose we define: A topological right-distributive quasigroup D will be called a quasigroup of Lie type if the group Σ generated by all the right translations of D is a Lie group.

We hope that this definition is no great restriction,

for it is rather natural and in fact we can prove that Σ is a Lie group in many cases. Some instances of this will be mentioned below:

(1) D is an almost-complex compact manifold, and the right translations of D preserve the almost-complex structure.

(2) D is a Riemannian manifold and the right translations are at the same time isometries of D.

(3) D is a manifold admitting an affine connection for which every right translation is an affine mapping.

(4) D is a C^1-differentiable compact manifold, and the automorphism group Σ which is (topologically) generated by the right translations possesses a neighbourhood of the identity consisting of equicontinuous automorphisms.

The main case in which I have solved the classification of right-distributive connected quasigroups of Lie type is the case of compact quasigroups. For these quasigroups, we have the following:

THEOREM. *Let* D *be a compact connected right-distributive quasigroup of Lie type*. *Then* D *is isomorphic to one of the following quasigroups*:

(1) *Let* T_n *be the n-dimensional torus group with* $n \geq 2$. *Let* α *be an automorphism of* T_n *such that the normalizer of the monothetic group* $\langle\bar{\alpha}\rangle$ *in the group* $\Sigma = T_n\langle\alpha\rangle$ *is* $\langle\bar{\alpha}\rangle$. (Since the continuous automorphism group of T_n is

discrete, we have $\langle\bar{\alpha}\rangle = \langle\alpha\rangle$.) Let $T_n(\alpha)$ be the class of conjugate elements of α in Σ. Then we get a two-sided distributive quasigroup on $T_n(\alpha)$ which is homeomorphic to T_n, where we define $x \circ y = y^{-1}xy$ for all $x, y \in T_n(\alpha)$. We shall denote these quasigroups by $T_n(\alpha)$. Two such quasigroups $T_n(\alpha_1)$ and $T_n(\alpha_2)$ are isomorphic exactly if there exists some isomorphism β from $T_n(\alpha_1)$ onto $T_n(\alpha_2)$ for which $\alpha_1^{\beta} = \alpha_2$.

(2) Let the Lie group Δ be the direct product of two Lie groups A and B with the following structure:

If $A \neq 1$, then A is the direct product of groups A_i with $i = 1, \ldots, n$, $n > 1$, such that A_1 is either the identity or a torus group of dimension at least two, and A_i for $i > 1$ is isomorphic to either $PSO_8(\mathbb{R})$ or $Spin_8(\mathbb{R})$. If $B \neq 1$, then B is the direct product of Lie groups B^k, where every B^k itself can be written as a direct product of groups B_j^k with amalgamated central subgroups isomorphic to the Klein four group. Here B_1^k either does not occur or, if it does, it is a two-dimensional torus group $SO_2 \times SO_2$; B_j^k for $j > 1$ appears to be isomorphic to $Spin_8(\mathbb{R})$.

Let Σ be the semidirect product of Δ with a discrete cyclic group ϕ generated by an automorphism α of Δ which acts on Δ in the following way:

α normalizes all the groups A_i and B_j^k; for $i, j > 1$, it acts on A_i and B_j^k as a triality automorphism (of order 3); finally, α induces on B_1^k (on A_1) an automorphism such that the normalizer of the cyclic group ϕ in $B_1^k\phi$ (in $A_i\phi$) is identical with ϕ.

Now let $D(\Delta, \alpha)$ be the conjugate class of α in Σ. Defining $x \cdot y = y^{-1}xy$ for $x, y \in D(\Delta, \alpha)$ yields a right-distributive quasigroup which is not left-distributive. This quasigroup is homeomorphic to the topological product

$$A_1 \times \prod_{i=2}^{n} (A_i/C_i) \times (B^k / \prod_{j=1}^{m} {}^kC_j^*)$$

where all the C_i and ${}^kC_j^*$ are homeomorphic to the compact Lie groups of exceptional type G_2.

Two quasigroups $D(\Delta_1, \alpha_1)$ and $D(\Delta_2, \alpha_2)$ are isomorphic if and only if there exists an isomorphism β of $\Delta_1\langle\alpha_1\rangle$ to $\Delta_2\langle\alpha_2\rangle$ for which $\alpha_1^{\beta} = \alpha_2$.

In our final classification theorem, we determine all those right-distributive quasigroups of Lie type which are homeomorphic to the plane $\mathbb{R}^2$, and which we therefore call the planar quasigroups. The classification of planar quasigroups appears to me to be justified because the only surfaces which admit a regular idempotent multiplication are the plane and the torus. We state

THEOREM. Let D be a planar right-distributive quasigroup of Lie type. Then D is isomorphic to one of the following quasigroups:

(a) Let Σ be the group of proper euclidean motions, and let α be a rotation through an irrational multiple of π.

Let $S(\alpha)$ be the class of conjugate elements in Σ. If we define $x \circ y = y^{-1}xy$ for $x, y \in S(\alpha)$, then $S(\alpha)$ is a two-sided distributive quasigroup.

The right translations of $S(\alpha)$ always generate Σ. Two quasigroups obtained in this fashion are isomorphic exactly if there exists a continuous automorphism γ of Σ with $\alpha_1^{\gamma} = \alpha_2$.

(b) Let α be an element of $GL_2(\mathbb{R})$ which generates a discrete subgroup in $GL_2(\mathbb{R})$. Let Σ_α be the semidirect product of the euclidean translation group $\mathbb{R}^2$ with $\langle\alpha\rangle$. Let $J(\alpha)$ be the class of conjugate elements of α in Σ. Then $J(\alpha)$ forms a planar two-sided distributive quasigroup, if we define a product $x \circ y = y^{-1}xy$, for $x, y \in J(\alpha)$. The group (topologically) generated by the right translations of $J(\alpha)$ is isomorphic to Σ.

Two quasigroups $J(\alpha_1)$ and $J(\alpha_2)$ are isomorphic exactly if there exists some isomorphism β of the group Σ_{α_1} onto Σ_{α_2} which maps α_1 to α_2.

(c) The group generated by the right translations of D is $PSL_2(\mathbb{R})$, which has been discussed earlier.

(d) Let Δ be the group of matrices

$$\left\{ g(t, u, v) = \begin{pmatrix} 1 & u & v \\ 0 & a^{-t} & 0 \\ 0 & 0 & a^{t} \end{pmatrix} \middle| \; u, v, t \in \mathbb{R}, a \geq 1 \text{ fixed} \right\}$$

Let α be the involution $\begin{pmatrix} 1 & 0 & 0 \\ 0 & 0 & 1 \\ 0 & 1 & 0 \end{pmatrix}$ and let Σ be the semidirect product of Δ with $\langle\alpha\rangle$. Then the class R of conjugate elements of α in Σ forms a planar right-distributive quasigroup if we take conjugation as multiplication.

In R, the left-distributive law is violated. All the quasigroups obtained in this manner are isomorphic because the isomorphy type of Δ is independent of the real number a in the definition of the matrices g(t, u, v).

(e) Let Δ be the nilpotent group of matrices

$$\left\{ g(t, u, v) = \begin{pmatrix} 1 & u & v \\ 0 & 1 & t \\ 0 & 0 & 1 \end{pmatrix}; \ u, r, t \in \mathbb{R} \right\}$$

which is homeomorphic to $\mathbb{R}^3$, and let α_p be the matrix

$$\begin{pmatrix} 1 & 0 & 0 \\ 0 & 1 & 0 \\ 0 & 0 & p \end{pmatrix}$$

with a real number $0 \neq p \neq 1$. We denote by Σ the semidirect product of Δ with the cyclic group $\langle\alpha_p\rangle$.

Then the class S_p of elements which are conjugate to α_p in Σ forms a planar two-sided distributive quasigroup if we set $x \cdot y = y^{-1}xy$ for all $x, y \in S_p$. Two two-sided quasigroups S_{p_1} and S_{p_2} defined in this manner are isomorphic exactly if $p_1 = p_2$.

REFERENCES

1. J. Aczel. Vorlcsungen über Funktionalgleichungen und ihre Anwendungen. Birkhäuser Verlag, Basel (1961).
2. A. Borel. Sur la cohomologie des espaces fibrés principaux et des espaces homogènes des groupes de Lie compacts. Annals of Math. 57 (1953): 115-207.
3. B. Fischer. Distributive Quasigruppen endlicher Ordnung. Math. Z. 83 (1964): 267-303.

4. H. Freudenthal. Oktaven, Ausnahmegruppen und Oktavengeometrien. (Vorlesungsausarbeitung) Utrecht 1960.
5. H. Hopf. Über die Topologie der Gruppenmannigfaltigkeiten und ihre Verallgemeinerungen. Annals of Math. 42 (1941): 22-52.
6. —— Fundamentalgruppe und die zweite Bettische Gruppe. Comm. Math. Helvet. 14 (1942).
7. H. Kneser. Sur les variétés connexes de dimension 1. Bull. Soc. Math. Belgique 10 (1958): 19-25.
8. H. Kneser and M. Kneser. Real analytische Strukturen der Alexandroff-Halbgeraden und der Alexandroff-Geraden. Archiv der Math. 11 (1960): 104-106.
9. T.A. Springer and F.D. Veldkamp. On Hjelmslev-Moufang planes. Math. Z. 107 (1968): 249-263.
10. K. Strambach. Rechtsdistributive Quasigruppen auf Mannigfaltigkeiten. Math. Z. (to appear).

PLANES WITH GIVEN GROUPS

Jill Yaqub

Let π be a projective plane, and let Δ be a collineation group of π. Then Δ may be "given" in one or more of the following senses: (1) Δ is isomorphic to a specific abstract group Δ_1 (e.g. "$\Delta \simeq PSL(3, q)$"), (2) Δ, or some subset of Δ, acts on π in a prescribed way (e.g. "Δ is transitive on the points of π," or "Δ is of Lenz class $\geq$ IV," or "Δ contains involutory perspectivities"), (3) Δ has a given abstract property (e.g. "Δ is abelian," or "Δ is solvable").

We shall be concerned mainly with finite planes, though in some cases it is sufficient to assume only that Δ is finite (see sections 3 and 4). If π is a plane "with given group Δ," it is assumed that Δ is faithful on π (i.e. that if $\delta \in \Delta$ fixes all points of π, then $\delta = 1$). We shall discuss examples of two types of problem: I, given that $\Delta \simeq \Delta_1$ (as in (1)), and possibly that there is some relation between the orders of π and Δ (e.g. "π is of order q^2 and $\Delta \simeq PSL(3, q)$"), determine (i) the structure of π, (ii) the possible actions of Δ on π; II, given that π admits Δ, where Δ has a partially prescribed action on π (as in (2)), determine (i) the structure of π, (ii) the possible actions of Δ on π, (iii) the abstract structure of Δ.

A problem of type I is usually solved as follows. (a) The known abstract properties of Δ are used to show that $\pi \simeq \pi(\Delta)$, where the points and lines of $\pi(\Delta)$ are specified objects within Δ (such as subgroups or cosets), and incidence is described entirely in terms of Δ. (b) An isomorphism is established between $\pi(\Delta)$ and $\pi'(\Delta')$, where π' is a known plane which admits $\Delta' \simeq \Delta$ acting in a known way. It then follows that $\pi \simeq \pi'$ and that Δ acts on π as Δ' acts on π'. This procedure is illustrated in some detail in section 1, where we describe the work of Dembowski [6], [7], [8] and Unkelbach [26] on planes of order q^2 which admit a collineation group $\Delta \simeq PSL(3, q)$.

A problem of type II can be reduced to a problem of type I if it can be shown that Δ (or some sufficiently large subgroup of Δ) is isomorphic to a specific abstract group. (This approach was first used by Lüneburg, 1964a, c.[1]) To achieve this, it is usually necessary to apply an appropriate deep group-theoretic characterization theorem to a homomorphic image Γ of some subgroup of Δ. Thus the feasibility of solving a type II problem by this method depends on the existence of the "right" characterization theorem.

1. References cited only by author and date can be found in the bibliography of Dembowski [5].

Even then, to show that Γ satisfies the hypotheses of the characterization theorem and to determine Δ from Γ may require a complicated group-theoretic analysis involving other deep results. In section 2 we outline the proof of a type I result due to Hoffer [17], and then show (for the easiest case) how Kantor [21] reduces a related type II problem to the problem solved by Hoffer.

It is usually very helpful to know that Δ contains perspectivities other than the identity. The collineation δ is a "(P, ℓ) perspectivity" if it fixes all points on the line ℓ and all lines on the point P; if $P \in \ell$, δ is an "elation," while if $P \notin \ell$, δ is a "homology." If $\delta \neq 1$, then P and ℓ are uniquely determined by δ, and are called its "centre" and "axis" respectively. For the basic properties of perspectivities, see Dembowski [5], §3.1. The existence of involutory perspectivities in a collineation group of even order can often be deduced from a fundamental theorem of Baer, 1946b.

DEFINITION. The subplane $\pi' \subseteq \pi$ is a "Baer subplane" of π <=> each point $x \in \pi - \pi'$ is on exactly one line of π', and dually.

Baer subplanes are "maximal," in the sense that if $\pi' \subseteq \pi'' \subseteq \pi$, where π'' is a subplane and π' a Baer subplane of π, then either $\pi'' = \pi'$ or $\pi'' = \pi$. It is easily shown that if π' is a Baer subplane of a plane π of finite order n, then $n = m^2$ and π' has order m.

THEOREM (Baer's Theorem on Involutions). Let π be an involutory collineation of π (i.e. $\sigma^2 = 1$ but $\sigma \neq 1$). Then either (i) σ is a (P, ℓ) perspectivity for some (P, ℓ), or (ii) the points and lines of π which are fixed by σ form a Baer subplane π' of π. If π is of finite order n, then in case (i), $P \in \ell \iff n$ is even, and in case (ii), $n = m^2$ and π' has order m.

An involution of type (ii) is called a "Baer involution." In studying problems of type I or II, with $|\Delta|$ even, it is usually the first objective to show that Δ contains involutory perspectivities, and that Δ, or at least some "large" subgroup of Δ, contains no Baer involution. An interesting feature of both proofs in section 2 is the way in which the authors use a theorem of Seib [25] to handle possible Baer involutions in Δ. (For more about the role of Baer involutions, see Kantor [22].)

In sections 3 and 4, we describe some recent results on finite collineation groups which contain non-trivial elations and non-trivial homologies respectively. Section 3 is centred on a paper by Hering [16] whose results are in a sense complementary to those of Piper, 1965, 1966a. In section 4 we state and briefly discuss results of Piper, 1967, Brown [3], [4], Hering [14], and Kantor [23].

Our starting-point is Chapter IV of Dembowski's

"Finite Geometries" [5], and we have systematically excluded any detailed discussion of work described there. We have tried to illustrate techniques by giving some parts of the proofs in detail, but there are some obvious omissions, particularly as regards arguments used in applying characterization theorems. For these, see, for example, Lüneburg 1964a, c, Hering and Kantor [20], Kantor [21], [23]. We have made no attempt to give a comprehensive survey of recent results; a few additional references on closely related topics are included in the bibliography. The following are suggested as basic references: Dembowski [5], Dickson, 1901, Feit [9], Gorenstein [12], Hall, 1959, Hughes and Piper [18], Huppert [19], Pickert, 1955, and Wielandt, 1964.

I am grateful to Professor W.M. Kantor for sending me a preprint of [23] and allowing me to describe its results before publication, to Professors C. Hering, H. Lüneburg, and S.K. Wong for helpful discussion, and to Dr. J. Sunday for assistance in preparing the manuscript.

NOTATION. Points and lines will be denoted by capital and lower case letters respectively. Collineation groups and their elements will usually be denoted by capital and lower case Greek letters respectively. If Σ is a subgroup of Δ, $C_\Delta(\Sigma) = \{\delta \in \Delta \mid \delta\sigma = \sigma\delta$ for all $\sigma \in \Sigma\}$ and $N_\Delta(\Sigma)$ $= \{\delta \in \Delta \mid \delta^{-1}\Sigma\sigma \subseteq \Sigma\}$ are the "centralizer" and "normalizer" of Σ in Δ respectively. The centre of the group Δ is denoted by $Z\Delta$ and the centre of the skew-field K by ZK. (Note that we regard the class of commutative fields as a subclass of the class of skew-fields - i.e. a "skew-field" may have commutative multiplication.) If Δ is a permutation group on the set S, then, for $x \in S$, $\Delta_x = \{\delta \in \Delta \mid \delta(x) = x\}$ is the "stabilizer" of x in Δ. As usual, $\langle\alpha, \beta, \ldots, \kappa\rangle$ denotes the group generated by $\alpha, \beta, \ldots, \kappa$. If p is a prime, we say $p^a \| n \iff p^a$ divides n but p^{a+1} does not divide n. If a, b are integers, then (a, b) denotes the greatest common divisor of a and b.

1. PSL(3, q) AND THE GENERALIZED HUGHES PLANES

A precise definition of the generalized Hughes planes will be given later. In the meantime we note the following distinction between "Hughes planes" and "generalized Hughes planes." "Hughes planes," as defined by Hughes, 1957b, are constructed from a near-field F which is of dimension 2 over its kernel K and which has the property ZF = K. It was pointed out by Ostrom, 1968, and Dembowski [7] that the restriction ZF = K is unnecessary for the construction. Thus for the "generalized Hughes planes" it is required only that F be of dimension 2 over K. (In particular, K need not be commutative in the infinite case.) The finite near-fields have been completely determined by Zassenhaus, 1935a. (See also [5], pp. 230-231.) The "regular" finite near-fields all satisfy $ZF = K$, and so do 3 of the 7 "irregular" near-fields. Thus there exist only 4 finite near-fields with $ZF \neq K$ (numbers I, III, V, and VI of [5], p. 231.) Their orders are 5^2, 7^2, 11^2, and 29^2, and their centres are of orders 2, 2, 2, and 14 respectively. We shall call these near-fields "special," and the associated generalized Hughes planes "special Hughes planes."

Each generalized Hughes plane H contains a desarguesian Baer subplane H_0 which is fixed (as a set) by all collineations of H. The full collineation group of H has just 2 point-orbits, namely the points of H_0 and the points of $H - H_0$, and dually has just 2 line-orbits. Every perspectivity of H_0 is induced by some perspectivity of H. Conversely, the following result was proved by Ostrom, 1965c. (For simplicity, we state the hypothesis in a slightly stronger form than is necessary.)

THEOREM 1.1. Let π be a plane which contains a desarguesian Baer subplane π_0. If every perspectivity of π_0 can be extended to a perspectivity of π, then π is either desarguesian or a generalized Hughes plane.

The proof is by coordinates. Ostrom raises the question of whether the theorem remains true if "every perspectivity" is replaced by "every elation." The object of this section is to describe the proof of the following related result.

THEOREM 1.2 (Dembowski [6], [7], [8], Unkelbach [26]). *Let* π *be a projective plane of finite order* q^2, *where* q *is a prime-power. If* π *admits a collineation group* $\Delta \simeq$ PSL(3, q), *then* π *is either desarguesian or a generalized Hughes plane*.

The term "generalized" does not appear in the theorems stated by Dembowski and Unkelbach. However, as we shall see, the special Hughes planes of order 5^2, 11^2, 29^2 do satisfy the hypotheses of the theorem. Professor Lüneburg tells me that the discrepancy has already been noticed by others, but, as far as I know, no correction has been published. The difficulty arises from an oversight in [8], (2.9).

We first recall the definition of the group PSL(3, q), and note some of its properties. Let $K = GF(q)$ be the finite field with q elements. Let GL(3, q) be the group of all non-singular 3×3 matrices with elements in K, and $SL(3, q) = \{A \in GL(3, q) \mid \det A = 1\}$. If I denotes the identity matrix and K^* the multiplicative group of K, let $\hat{K} = \{kI \mid k \in K^*\}$. Then, by definition, $PGL(3, q) = GL(3, q)/\hat{K}$ and $PSL(3, q) = SL(3, q)/[SL(3, q) \cap K]$. Now $SL(3, q) \cap \hat{K} = \{kI \mid k^3 = 1\}$, so that $SL(3, q) \cap \hat{K} = \{1\}$ if $q \not\equiv 1 \pmod 3$, and otherwise has order 3. It follows easily from this and the fact that the map $\phi: GL(3, q) \to K$ with $\phi(A) = \det A$ is an epimorphism that $|PGL(3, q)| =$

$|SL(3, q)| = q^3(q - 1)^2(q + 1)(q^2 + q + 1)$, and that $|PSL(3, q)| = |SL(3, q)|/d$, where $d = (3, q - 1)$. Thus if $q \not\equiv 1 \pmod 3$, then $SL(3, q) \simeq PGL(3, q) = PSL(3, q)$; if $q \equiv 1 \pmod 3$, then PSL(3, q) is a proper subgroup of PGL(3, q) and a proper factor group of SL(3, q). Geometrically, PGL(3, q) is the projective group of the desarguesian plane π_q of order q (i.e. the group generated by all perspectivities of π_q), and PSL(3, q) is the little projective group of π_q (i.e. the group generated by all elations of π_q). Note that PSL(3, q) is always simple. (See Huppert [19], p.182.)

The principal results of Dembowski [6] are given in Theorems 1.3 and 1.4.

THEOREM 1.3. *Let* π *be of order* q^2, *and let* π_0 *be a Baer subplane of* π. *Suppose that* Δ *is a collineation group of* π *which fixes* π_0 *as a set and is faithful on* π_0. *Then* Δ *is flag-transitive on* $\pi - \pi_0 \iff \Delta$ *contains a subgroup* $\simeq PSL(3, q)$.

THEOREM 1.4. *Let* π *be of order* q^2, *and let* π_0 *be a desarguesian Baer subplane of* π. *Let* Δ *be a collineation group of* π *which acts faithfully on* π_0 *and contains the little projective group of* π_0. *Then* (a) *if* $\lambda \in \Delta$ *induces an elation of* π_0, λ *is an elation of* π; (b) *if* $1 \neq \lambda \in \Delta$ *induces a homology of* π_0, *then all fixed points of* λ *in* $\pi - \pi_0$ *are on the axis of* λ.

PROOF OF THEOREM 1.4(a). Let $\ell \in \pi_0$, and let Λ be the subgroup of Δ consisting of those collineations which induce elations of π_0 with axis ℓ. Since Δ is faithful on π_0 and contains the little projective group of π_0, $|\Lambda| = q^2$; moreover, all collineations $\neq 1$ in Λ are conjugate in Δ. Let $S = \ell - (\ell \cap \pi_0)$; then $|S| = (q^2 + 1) - (q + 1) = q(q - 1)$. Each $\lambda \in \Lambda$ with $\lambda \neq 1$ must fix the same number of points in S: denote this number by f. It suffices to show that $f \geq q(q - 1)$. Let x be the number of orbits of Λ on S. For each such orbit O, the total number of points in O fixed by the elements of Λ is $|O||\Lambda_X|$, where X is an arbitrary point of O. However, $|O||\Lambda_X| = |\Lambda|$. Hence, summing over the orbits O, we have

$$x|\Lambda| = |S| + \sum_{\substack{\lambda\in\Lambda,\\ \lambda\neq 1}} f,$$

i.e. $xq^2 = q(q - 1) + f(q^2 - 1)$. It follows that $f = qf'$, where $xq = (q - 1)[1 + f'(q + 1)]$. But then q divides $(1 + f')$, whence $f' \geq q - 1$ and $f \geq q(q - 1)$.

PROOF OF THEOREM 1.3. First suppose that Δ is flag-transitive on $\pi - \pi_0$ (i.e. transitive on incident point-line pairs (P, ℓ) with $P, \ell \in \pi - \pi_0$). Since π_0 is a Baer subplane of π, each such flag (P, ℓ) corresponds to a unique point-line pair $(P_0, \ell_0) \in \pi_0$, where P_0 is the unique point of ℓ which lies in π_0, and dually; clearly $P_0 \notin \ell_0$. Thus Δ is transitive on the non-incident pairs (P_0, ℓ_0) of π_0.

The result now follows from a theorem of Ostrom, 1958a.

Suppose, conversely, that Δ contains a subgroup $\Gamma \simeq PSL(3, q)$. Then, by Dembowski, 1965c, π_0 is desarguesian and Γ acts as the little projective group of π_0. Thus we may apply Theorem 1.4. Let $X, Y, X \cup Y \in \pi - \pi_0$. Then X is on a unique line x of π_0, Y is on a unique line y of π_0, and $X \cup Y$ is on a unique point C of π_0. Let $x \cap y = S$. Then $C \neq S$, and the $(C, S \cup C)$ elation which maps x on y maps X on Y. If $X, Z \in \pi - \pi_0$ but $X \cup Z \in \pi_0$, find $Y \in \pi - \pi_0$ such that $X \cup Y, Y \cup Z \notin \pi_0$ and apply the above argument twice. Then we see that Δ is transitive on the points of $\pi - \pi_0$. Thus it suffices to show that for $X \in \pi - \pi_0$, Δ_X is transitive on lines m such that $X \in m$ and $m \notin \pi_0$. Now X is on a unique line x of π_0, and each m is on a unique point M of π_0, where $M \notin x$. The group of all elations of π_0 with axis x is transitive on the points $M \in \pi_0$ with $M \notin x$, and by Theorem 1.4, all such elations fix X.

We omit the proof of Theorem 1.4(b), which involves a more complicated combinatorial analysis of the action on π of collineations which induce perspectivities in π_0.

We now summarize the main definitions and results of Dembowski [7], [8]. Let F be a right near-field of dimension 2 over the skew-field K. Then, by definition, F has all the properties of a skew-field except possibly the left distributive law $a(b + c) = ab + ac$, and F is a 2-dimensional left vector-space over K. (Note that F may itself be a skew-field. If F is not a skew-field, then K is the "kernel" of F, i.e. $K = \{k \in F \mid k(b + c) = kb + kc$ for all $b, c \in F\}$.) Let $e \in F - K$, and represent F on the basis $\{1, e\}$, so that each $f \in F$ is written uniquely as $f = x + ye$ with $x, y \in K$. For $x, y \in K$, define $e(x + ye) = u(x, y) + v(x, y)e$. Then u, v are "near-field functions of F over K." It is not hard to verify that u, v satisfy the following conditions (1) to (4).

(1) $(x, y) \to (u(x, y), v(x, y))$ is a permutation of $K \times K$.

(2) $u(1, 0) = 0, \quad v(1, 0) = 1, \quad u(0, 0) = v(0, 0) = 0.$

(3) $u(x_1x_2 + y_1u(x_2, y_2), x_1y_2 + y_1v(x_2, y_2))$
$= u(x_1, y_1)x_2 + v(x_1, y_1)u(x_2, y_2).$

(4) $v(x_1x_2 + y_1u(x_2, y_2), x_1y_2 + y_1v(x_2, y_2))$
$= u(x_1, y_1)y_2 + v(x_1, y_1)v(x_2, y_2).$

(To prove (3) and (4), expand the identity

$$[e(x_1 + y_1e)](x_2 + y_2e) = e[(x_1 + y_1e)(x_2 + y_2e)].$$

Note that (3) and (4) correspond to the fact that the set of matrices of the form $\begin{bmatrix} x & y \\ u(x,y) & v(x,y) \end{bmatrix}$ is closed under multiplication.) Conversely, Dembowski proves the following

THEOREM 1.5. *Given a pair of maps* u, v: $K \times K \to K$ *which satisfy conditions* (1) *to* (4), *there exists a right near-field* F *of dimension* 2 *over* K *such that, for some basis* $\{1, e\}$, u, v *are near-field functions of* F *over* K.

PROOF OUTLINE. Define multiplication on $F = K \oplus K$ by $(x_1, y_1)(x_2, y_2) = (x_1x_2 + y_1u(x_2, y_2), x_1y_2 + y_1v(x_2, y_2))$, and verify that F is a right near-field.

We shall now define the generalized Hughes planes. Let F be a right near-field of dimension 2 over the skew-field K. Let $W = F \oplus F \oplus F$, $V = K \oplus K \oplus K$. For $\underline{x} = (x_1, x_2, x_3) \in W$, define $\underline{x}f = (x_1f, x_2f, x_3f)$ for each $f \in F$. For $\underline{x}, \underline{y} \in W - \{\underline{0}\}$, define $\underline{x} \sim \underline{y} \iff \underline{x} = \underline{y}f$ for some $f \in F$. Then "$\sim$" is clearly an equivalence relation; denote the equivalence class which contains $\underline{x}$ by $\underline{x}F$. If $f \in F$ and $\underline{x} \in W$, let $f^*(\underline{x}) = x_1 + fx_2 + x_3$, and let $L(f) = \{\underline{x} \in W - \{\underline{0}\} \mid f^*(\underline{x}) = 0\}$. Clearly $\underline{x} \in L(f) \Rightarrow \underline{x}F \subseteq L(f)$, by the right distributive law in F. If $A \in GL(3, K)$ (the group of all non-singular 3×3 matrices over K), define $L(f, A) = \{A\underline{x} \mid \underline{x} \in L(f)\}$. Then $\underline{x} \in L(f, A) \Rightarrow \underline{x}F \in L(f, A)$. (We write $\underline{x}$ as a row-vector for typographical convenience, but it should in fact be regarded as a column-vector.) Let $\mathcal{P} = \{\underline{x}F \mid \underline{x} \in W - \{\underline{0}\}\}$, $\mathcal{L} = \{L(f, A \mid f \in F \text{ and } A \in GL(3, K)\}$. Let $H = H(F, K)$ be the incidence-structure $(\mathcal{P}, \mathcal{L}, G)$. Then:

THEOREM 1.6. (i) H <u>is a projective plane</u>. <u>It contains a Baer subplane</u> H_0 <u>whose points are the</u> $\underline{x}F$ <u>with</u> $\underline{x} \in V - \{\underline{0}\}$ <u>and whose lines are the</u> $L(f, A)$ <u>with</u> $f \in K$. (ii) H <u>admits a collineation group</u> $\Gamma \simeq GL(3, K)$ <u>which fixes</u> H_0 <u>as a set</u>; <u>in general</u> Γ <u>is not faithful on</u> H. (III) <u>The kernel of the representation of</u> Γ <u>on</u> H <u>is</u> $N = \{\lambda I \mid \lambda \in K^* \cap ZF\}$.

(iv) The kernel of the representation of Γ on H_0 is $N_0 = \{\lambda I \mid \lambda \in ZK^*\}$. (v) Γ is flag-transitive on $H - H_0$, and $H - H_0 \simeq (GL(3, K), A, B)$,* where A, B are the subgroups of $GL(3, K)$ which contain all matrices of the form

$$\begin{bmatrix} x & a_{12} & y \\ 0 & a_{22} & 0 \\ u(x,y) & a_{32} & v(x,y) \end{bmatrix}, \quad \begin{bmatrix} x & 0 & y \\ b_{21} & b_{22} & b_{23} \\ u(x,y) & 0 & v(x,y) \end{bmatrix}$$

respectively; here $u(x, y)$, $v(x, y)$ are near-field functions of F over K with respect to some basis $\{1, e\}$.

For the proof of Theorem 1.6, which involves a fair amount of calculation, see [7]. The plane $H = H(F, K)$ is a "generalized Hughes plane" provided that F is not a skew-field; if F is a skew-field, $H(F, K)$ is desarguesian. The group $\Gamma \simeq GL(3, K)$ acts as follows. Let $C \in GF(3, K)$. Then the point and line maps $\underline{x}F \to C(\underline{x}F)$ and $L(f, A) \to L(f, CA)$ are clearly bijective and preserve incidence, hence induce a collineation of H.

The representation of $H - H_0$ given in (v) is a "Higman-McLaughlin representation" (Higman and McLaughlin, 1961). If a group Γ acts on an incidence-structure $I = (P, L, \epsilon)$ (not necessarily faithfully), and is transitive on the flags of I, let $A = \Gamma_X$, $B = \Gamma_\ell$, where (X, ℓ) is a flag of I (i.e. $X \in P$, $\ell \in L$, and $X \in \ell$). Then A and B are determined to within conjugacy in Γ. Define the incidence structure J as follows. "Points" are the left-cosets ξA of A in Γ,

*This notation is explained below.

"lines" are the left-cosets $\eta\mathcal{B}$ of $\mathcal{B}$ in Γ, and $\xi\mathcal{A}$ is on $\eta\mathcal{B} \Longleftrightarrow \xi\mathcal{A} \cap \eta\mathcal{B}$ is not empty. Then J is isomorphic to I, and Γ acts on J by left multiplication. We write J = $(\Gamma, \mathcal{A}, \mathcal{B})$.

Let K = GF(q) and consider Theorem 1.6 (iii), (iv). If ZF = K, then $N = N_0$ and Γ induces the collineation group $\Gamma/N \simeq PGL(3, q)$ on H, where Γ/N is faithful on H_0. In [8], (2.9), Dembowski states that H admits a collineation group $\simeq PGL(3, q)$ acting faithfully on H_0 only if ZF = K. It is true that such a group is induced by Γ only if ZF = K. However, Γ contains a subgroup $\Sigma \simeq SL(3, q)$, and for the special near-fields of order q^2 with $q \not\equiv 1 \pmod 3$ (i.e. for q = 5, 11, 29), $\Sigma \cap N = \Sigma \cap N_0 = \{1\}$. Thus Σ is faithful on H and on H_0. Moreover, as noted earlier, for $q \not\equiv 1 \pmod 3$, $SL(3, q) \simeq PGL(3, q) = PSL(3, q)$. Hence the special Hughes planes of orders 5^2, 11^2, 29^2 satisfy the hypotheses of Theorem 1.2 and must be accounted for. In the case $q \equiv 1 \pmod 3$, Σ induces PSL(3, q) on H, acting faithfully on H_0, provided that ZF = K. For the special Hughes plane of order 7^2, Σ induces SL(3, 7) on H but PSL(3, 7) on H_0. Thus it is fairly clear that in this case H does not admit a collineation group $\simeq PSL(3, 7)$, and, as we shall see, Unkelbach's proof excludes the possibility.

In view of this difficulty, we shall modify the final stages of the proof of the main theorem in [8], by working in Σ instead of Γ. We shall prove Dembowski's result only in the case $q \not\equiv 1 \pmod 3$, since the case $q \equiv 1 \pmod 3$ is handled in Unkelbach's paper [26]. For $q \not\equiv 1 \pmod 3$, we have just seen that Σ induces on H a collineation group $\simeq PSL(3, q)$ which is faithful on H_0. Thus, by Theorem 1.3, Σ is flag-transitive on $H - H_0$. It follows that, for our purposes, Theorem 1.6(v) can be replaced by Theorem 1.6(v)'.

THEOREM 1.6(v)'. <u>If</u> $K = GF(q)$ <u>with</u> $q \not\equiv 1 \pmod 3$, <u>then</u> $H - H_0 \sim (\Sigma_1, \Sigma_1 \cap A, \Sigma_1 \cap B)$, <u>where</u> $\Sigma_1 = SL(3, q)$ <u>and</u> A, B <u>are the groups given in Theorem</u> 1.6(v).

THEOREM 1.7 (adapted from [8]). <u>Let</u> π <u>be of order</u> q^2, <u>where</u> q <u>is a prime-power</u> $\not\equiv 1 \pmod 3$. <u>Let</u> π_0 <u>be a Baer subplane of</u> π, <u>and suppose that</u> π <u>admits the collineation group</u> $\Delta \simeq PSL(3, q)$, <u>which fixes</u> π_0 <u>as a set</u>. <u>Then</u> π <u>is either desarguesian or a generalized Hughes plane</u>.

PROOF. (1) π_0 is desarguesian, and Δ acts as its little projective group.

If Δ fixes π_0 pointwise, it must be semiregular on the points of $\pi - \pi_0$, since π_0 is maximal. But then $|\Delta|$ must divide $q^4 - q$, which is not the case. Hence, since $PSL(3, q)$ is simple, Δ is faithful on π_0. The result now follows from Dembowski, 1965c.

(2) Δ is sharply transitive on the set of ordered pairs of lines in $\pi - \pi_0$ which intersect in $\pi - \pi_0$ (and dually).

The number of such ordered pairs of lines is $(q^4 - q)q^2(q^2 - 1)$, and this is the order of Δ. Hence it suffices to show that if $\delta \in \Delta$ fixes such a line pair, then $\delta = 1$. Suppose that δ fixes $g, h \in \pi - \pi_0$, where $X = g \cap h \in \pi - \pi_0$. Let G, H be the unique points of π_0 which are on g, h respectively and let x be the unique line of π_0 on X. Let $Y = (G \cup H) \cap x$. Then G, H, Y are 3 distinct collinear points which must be fixed by δ. Since δ belongs to the little projective group of the desarguesian plane π_0, it follows that δ induces a perspectivity of π_0, with axis $\ell = G \cup H$. But δ also fixes $X \notin \pi_0, \notin \ell$. Hence $\delta = 1$ by Theorem 1.4.

(3) Δ is flag-transitive on $\pi - \pi_0$, and if (P, ℓ) is a flag of $\pi - \pi_0$, then $\pi - \pi_0 \simeq (\Delta, \Delta_P, \Delta_\ell)$. This follows from (2) (or Theorem 1.3), and the theorem of Higman and McLaughlin, 1961, already mentioned.

(4) Let (P, ℓ) be a flag in $\pi - \pi_0$, let g be the unique line of π_0 on P and Q the unique point of π_0 on ℓ. Then Δ_P, Δ_ℓ are Frobenius groups, whose kernels are the group of all elations with axis g and the group of all elations with centre Q respectively. The group $\Delta_P \cap \Delta_\ell$ is a common Frobenius complement of Δ_P, Δ_ℓ and is sharply transitive on points $\neq Q, \notin g$ in π_0 and on lines $\neq g, \notin Q$ in π_0.

It follows from (2) that Δ_P is sharply 2-transitive on the set S of points of π_0 which are not on g, so that in particular Δ_P acts as a Frobenius group on S (i.e. Δ_P is transitive on S and only the identity fixes more than one point). The group of all elations in Δ with axis g is a normal subgroup of Δ_P which is fixed-point-free on S, and hence is the Frobenius kernel of Δ_P. Dually for Δ_ℓ. Since $\Delta_P \cap \Delta_\ell$ is the subgroup of Δ_P which fixes $Q \in S$, it is a Frobenius complement of Δ_P, and, dually, of Δ_ℓ.

(5) π is uniquely determined by $\pi - \pi_0$. See [5], p.317.

(6) Let K = GF(q), $q \not\equiv 1 \pmod 3$, and let $\bar{K} = K \times K - \{(0, 0)\}$. Let π_0 be the desarguesian plane over K, represented by homogeneous point and line coordinates $\underline{x} = (x, y, z)$, $\underline{\ell} = [\ell, m, n]$ respectively. Suppose that θ is a group of collineations of π_0, represented by transformations $\underline{x} \to A\underline{x}$ with $A \in SL(3, q)$, and that θ fixes the point Q = (0, 1, 0) and the line g = [0, 1, 0] and is sharply transitive on the set L of lines $\neq g$, $\not\ni Q$. Then

$$\theta = \left\{ \begin{bmatrix} x & 0 & y \\ 0 & a_{22} & 0 \\ u(x, y) & 0 & v(x, y) \end{bmatrix} \in SL(3, q) \;\middle|\; (x, y) \in \bar{K} \right\},$$

where u, v are near-field functions for some F over K, and a_{22} is determined by the fact that the matrix has determinant 1.

First consider θ as a group of line-transformations $\underline{\ell} \to \underline{\ell}M$, where $M \in SL(3, q)$. Since Q and g are fixed,

$$M = k \begin{bmatrix} a & 0 & b \\ 0 & 1 & 0 \\ c & 0 & d \end{bmatrix}, \quad \text{with } k^3(ad - bc) = 1.$$

Then $[1, 1, 0]M = k[a, 1, b]$ and $[0, 1, 1]M = k[c, 1, d]$. Since θ is sharply transitive on $L = \{[\ell, 1, n] \mid (\ell, n) \in \bar{K}\}$, we see that $\{(a, b)\} = \bar{K}$ and that the map $\phi: \bar{K} \to \bar{K}$ with $\phi(a, b) = (c, d)$ is bijective. It is easily verified that c, d, as functions of (a, b), satisfy (1) to (4) of Theorem 1.5 (where we define $c(0, 0) = d(0, 0) = 0$). Hence there exists a near-field F of dimension 2 over K, and an element $e \in F - K$ such that $e(a + be) = c + de$ for all $M \in \theta$.

The line-transformation $\underline{\ell} \to \underline{\ell}^T M$ is represented in point-coordinates by $\underline{x} \to A\underline{x}$, where

$$A = (M^{-1})^T = k^2 \begin{bmatrix} d & 0 & -c \\ 0 & k^{-3} & 0 \\ -b & 0 & a \end{bmatrix}.$$

Let $ee' = -1$. Then $\{1, e'\}$ is also a basis for F over K, and since $-1 \in ZF$, $e'e = -1$. Since $e[(a + be)e'] = (c + de)e'$ and since F is right distributive, $e(ae' - b) = ce' - d$, whence multiplying on the left by $-e'$, $-b + ae' = e'(d - ce')$. Thus $-b$, a are near-field functions of d, $-c$ with respect to the basis $\{1, e'\}$. If F is a special near-field of order 5^2, 11^2, or 29^2, then $ad - bc \in ZF$ for any near-field functions c, d of a, b: this can be verified by inspection for the representations given in [5], p.231, and it can be shown that, for a given $f \in F$, $ad - bc$ is independent of the choice of the basis element e. Since here $k^3(ad - bc) = 1$, since K has cyclic multiplication, and since $q \not\equiv 1 \pmod 3$, it follows that $k \in ZF$. If F is not special, then certainly $k \in ZF$. But then $e'(k^2d - k^2ce') = k^2e'(d - ce') = k^2(-b) + k^2ae'$. Hence

$$A = \begin{vmatrix} x & 0 & y \\ 0 & a_{22} & 0 \\ u(x, y) & 0 & v(x, y) \end{vmatrix},$$

where u, v are near-field functions of (x, y) and det A = 1. This implies that A is uniquely determined by its first row (given the basis {1, e'}), where a_{22} is determined by the fact that det A = 1. Since $|\theta| = |\bar{K}|$, $\{(x, y) \mid (x, 0, y)$ is the first row of $A \in \theta\} = \bar{K}$.

(7) Let $\Sigma_1 = SL(3, q)$, and let $\mathcal{A}$, $\mathcal{B}$ be defined as in Theorem 1.6. Let (P, ℓ) be a flag of $\pi - \pi_0$. Then $(\Delta, \Delta_P, \Delta_\ell) \simeq (\Sigma_1, \Sigma_1 \cap \mathcal{A}, \Sigma_1 \cap \mathcal{B})$.

Let g be the unique line of π_0 on P and Q the unique point of π_0 on ℓ. Choose homogeneous coordinates in π_0 so that $Q = (0, 1, 0)$ and $g = [0, 1, 0]$. Since $\Delta \simeq PSL(3, q)$ and $q \not\equiv 1 \pmod 3$, $\Delta \simeq \Sigma_1$, and we can represent Δ as the group of all point-transformations $\underline{x} \to A\underline{x}$ of π_0 such that $A \in \Sigma_1$. By (4), $\Delta_P = \Lambda\theta$, where Λ is the group of all elations in Δ with axis g, and θ satisfies the hypotheses of (6). Since Λ is represented by the group of all matrices of the form

$$\begin{bmatrix} 1 & a_{12} & 0 \\ 0 & 1 & 0 \\ 0 & a_{13} & 1 \end{bmatrix},$$

Δ_P is represented by the group of all matrices of he form $\Sigma_1 \cap \mathcal{A}$, i.e. $\Delta_P \simeq \Sigma_1 \cap \mathcal{A}$. Dually $\Sigma_\ell \simeq \Sigma_1 \cap \mathcal{B}$.

Finally, by (5), (7), and Theorem 1.6 (v)', π is desarguesian or a generalized Hughes plane.

REMARK. The reader may wonder why we use line-coordinates in (6). If we try to determine the form of A directly from the (equivalent) hypothesis that θ is sharply transitive on points $\neq Q$, $\not\in g$ in π_0, we obtain near-field functions acting on the columns, not rows, of A. The present form of A is required for the comparison with Theorem 1.6(v)'. The corresponding part of Dembowski's proof ([8], (2.5), (3.7)) is simpler, partly because he is assuming ZF = K and partly because he uses the fact that the (ordinary) Hughes planes are known to be self-dual (Rosati, 1960a). It follows from (7) and the proof of (6) that the special Hughes planes of orders 5^2, 11^2, 29^2 are also self-dual. For the map $A \to (A^{-1})^T$ is an automorphism of $\Sigma_1 = SL(3, q)$, which interchanges $\Sigma_1 \cap \mathcal{A}$ and $\Sigma_1 \cap \mathcal{B}$, and hence induces a duality of $(\Sigma, \Sigma_1 \cap \mathcal{A}, \Sigma_1 \cap \mathcal{B})$.

We now turn to the paper by Unkelbach [26]. Its principal results are the following:

THEOREM 1.8. _If_ π _is of order_ q^2 _and admits a collineation_ group $\Delta \simeq PSL(3, q)$, then Δ fixes a Baer subplane π_0 of π. Moreover Δ is faithful on π_0.

THEOREM 1.9. *If π is of order q^2, where q is a prime-power $\equiv 1 \pmod 3$, and if π admits a collineation group $\Delta \simeq PSL(3, q)$, then π is either desarguesian or a Hughes plane.*

We shall describe the proofs only for the case q odd. (The proof of Theorem 1.8 is slightly harder in the case $q = 2^n$, but the proof of Theorem 1.9 is much easier.)

Let $\Delta \simeq PSL(3, q)$. For q odd, the number of involutions in Δ is $q^2(q^2 + q + 1)$, and they are all conjugate in Δ. If σ is an involution in Δ, $C_\Delta(\sigma) \simeq GL(2, q)/Z_0$, where Z_0 is the cyclic subgroup of $Z(GL(2, q))$ with order $(3, q - 1)$. Besides σ, $C_\Delta(\sigma)$ contains $q^2 + q$ involutions $\neq \sigma$, all of which are conjugate in $C_\Delta(\sigma)$. $C_\Delta(\sigma)$ contains a subgroup $\simeq SL(2, q)$ whose only involution is σ. (These facts can easily be verified by considering PSL(3, q) as the little projective group of the desarguesian plane of order q: in this representation the involutions are homologies.) If σ is a (P, ℓ) homology, then $C_\Delta(\sigma)$ is the subgroup fixing P and ℓ. $C_\Delta(\sigma)$ contains the group generated by all (X, XP) elations with $X \in \ell$, and this is isomorphic to SL(2, q). $|SL(2, q)| = q(q^2 - 1)$. (See Huppert [19], p.178.)

PROOF OF THEOREM 1.8 (q odd)

(1) The involutions in Δ are homologies.

Suppose $\sigma \in \Delta$ is a Baer involution with fixed subplane π'. Let $\Sigma \subseteq C_\Delta(\sigma)$, where Σ is isomorphic to a Sylow 2-group of SL(2, q). Let $|\Sigma| = 2^a$, so that $2^a \| (q - 1)(q + 1)$. Let $\alpha \in \Sigma$, $\alpha \neq \sigma, 1$. If α fixes $X \notin \pi'$, then σ fixes X, since σ, the unique involution in Σ, must be a power of α. Hence Σ is semiregular on the points of $\pi - \pi'$. But then $2^a \| [(q^4 + q^2 + 1) - (q^2 + q + 1)]$, i.e. $2^a \| q(q - 1) \times (q^2 + q + 1)$, i.e. $2^a \| (q - 1)$, a contradiction.

(2) If $\alpha \in \Delta$ is an involutory (P, g) homology and if $\beta \neq \alpha$ is an involution in $C_\Delta(\alpha)$, then β is an (X, y) homology where $X \neq P$, $y \neq g$, $X \in g$, and $P \in y$. Moreover β is the only (X, y) involutory homology.

Suppose β is a (P, g) homology in $C_\Delta(\alpha)$, $\beta \neq \alpha$. Then the $q^2 + q$ involutory homologies $\neq \alpha$ in $C_\Delta(\alpha)$ all have centre P and axis y, since they are all conjugate in $C_\Delta(\alpha)$. But there exist at most $(q^2 - 1)$ distinct (P, g) homologies, since π has order q^2. Since β and α commute, (2) now follows from a theorem of Ostrom and Lüneburg ([5], p.120; see also section 4 of these notes).

(3) Δ has no fixed point or line.

Suppose Δ fixes the point P. If one involution in Δ has centre P, then so do all such involutions, since all are conjugate in Δ. This contradicts (2). So P must be on the axis of each involutory homology in Δ. Let α be an involutory (R, g) homology; then $P \in g$. If β is an involution in $C_\Delta(\alpha)$, then, by (2), β has centre Q and axis PR, where $Q \in g$. But then $\alpha\beta$ is an involutory homology with centre P. Hence Δ fixes no point, and dually no line.

(4) There exists a line which is the axis of involutory homologies in Δ with different centres and which also contains the centre of an involutory homology in Δ.

Let α be an involutory (P, g) homology in Δ. By (2), the $q^2 + q$ involutions in $C_\Delta(\alpha)$ all have centre on g and axis on P, and no centre-axis pair yields more than one such involution. Since there are only $q^2 + 1$ lines on P, some axis h must have more than one centre. Then h satisfies the required conditions.

By (3), (4), and a theorem of Piper, 1967, the centres and axes of homologies in Δ form a desarguesian subplane π_0, and Δ restricted to π_0 contains the little projective group of π_0. Since there are $q^2(q^2 + q + 1)$ involutions in Δ, each corresponding to a unique non-incident point-line pair, and since Δ is transitive on the non-incident point-line pairs of π_0, π_0 has order q. It follows that Δ is faithful on π_0, and acts as its little projective group.

PROOF OUTLINE FOR THEOREM 1.9 (q odd). Suppose that π is of order q^2, where q is an odd prime power $\equiv 1 \pmod 3$, and that π admits a collineation group $\Delta \simeq PSL(3, q)$. Then Δ acts on π as in Theorem 1.8.

(1) Δ is flag-transitive on $\pi - \pi_0$, and $\pi - \pi_0 \simeq$

$(\Delta, \Delta_P, \Delta_\ell)$, where (P, ℓ) is a flag in $\pi - \pi_0$. Δ_P, Δ_ℓ are Frobenius groups of orders $q^2(q^2 - 1)/3$, and $\Delta_{P,\ell} = \Delta_P \cap \Delta_\ell$ is a common Frobenius complement for Δ_P, Δ_ℓ. If g is the unique line of π_0 on P, then $\Delta_P = \Lambda \cdot \Delta_{P,\ell}$, where Λ is the group of all elations in Δ with axis g (and dually). The Sylow p-groups of $\Delta_{P,\ell}$ are cyclic if p is odd, and the Sylow 2-groups are cyclic or generalized quaternion.

The last statement follows from the fact that $\Delta_{P,\ell}$ is a Frobenius complement and from a theorem of Burnside (see Huppert (19], p.499). The remaining statements are proved as in Theorem 1.7, (4). Note that we may apply Theorems 1.3 and 1.4; in particular, all elations of π_0 are elations of π. Theorem 1.4(b) is used once again to show that no element $\neq 1$ in Δ fixes 2 lines of $\pi - \pi_0$ which intersect in $\pi - \pi_0$, whence it follows that Δ_P is a Frobenius group on the set S of points of π_0 which are not on g. The principal difference between the situations of Theorems 1.7 and 1.9 is that Δ, hence $\Delta_{P,\ell}$, is now one-third the size it was before; in particular, Δ_P is no longer 2-transitive on S.

Henceforth let (P, ℓ) be a flag of $\pi - \pi_0$, let g be the unique line of π_0 on P, and let Q be the unique point of π_0 on ℓ. Let σ be the (unique) involutory (Q, g) homology. Then $\Delta_{Q,g} = C_\Delta(\sigma)$ contains a group $\Sigma \simeq SL(2, q)$. Let $\bar{g} = g - (g \cap \pi_0)$, and let $R \in \bar{g}$.

(2) Σ can act on $\bar{g}$ as follows: (a) Σ is transitive

on $\bar{g}$ and Σ_R is cyclic of order q + 1; (b) Σ has 2 orbits of length q(q - 1)/2, and $\Sigma_R = N_\Sigma(Z)$, where Z is a cyclic group of order q + 1 and $|\Sigma_R| = 2(q + 1)$; (c) $q \equiv 3 \pmod 4$, Σ is transitive on $\bar{g}$, $|\Sigma_R| = q + 1$, and Σ_R contains a cyclic subgroup of index 2. (In cases (b) and (c), Σ_R is non-abelian.)

Σ has no fixed point on $\bar{g}$, since it is generated by the (X, XQ) elations of π_0 with $X \in g \cap \pi_0$: these are elations of π, by Theorem 1.4(a). The involution σ fixes $\bar{g}$ pointwise, where Σ induces $\Gamma \simeq \Sigma/\langle\sigma\rangle \simeq PSL(2, q)$ on $\bar{g}$. (Since $q \equiv 1 \pmod 3$ we can assume $q \geq 7$, so that PSL(2, q) is simple.) If $R \in \bar{g}$, then for a given $X \in g \cap \pi_0$, the orbit of R under the group of (X, QX) elations is of length q. Thus $q \leq |R^\Gamma| \leq q(q - 1)$. Since $q(q^2 - 1)/2 = |\Gamma| = |\Gamma_R||R^\Gamma|$,

(*) $(q + 1)/2 \leq |\Gamma_R| \leq (q^2 - 1)/2.$

The subgroups of PSL(2, q) were completely determined by Dickson, 1901 (see also [19], p.213). The only candidates for Γ_R which satisfy the inequality (*) are the following: (i) Γ_R is cyclic of order (q + 1)/2; (ii) Γ_R is the normalizer in Γ of a cyclic group of order (q + 1)/2 and $|\Gamma_R| = q + 1$; (iii) Γ_R is the normalizer in Γ of a cyclic group of order (q - 1)/2 and $|\Gamma_R| = q - 1$; (iv) $q \equiv 3 \pmod 4$, and Γ_R is dihedral of order (q + 1)/2; (v) $\Gamma_R \simeq A_4$; (vi) $\Gamma_R \simeq S_4$ and $q^2 \equiv 1 \pmod{16}$; (vii) $\Gamma_R \simeq A_5$ and either q = 5 or $q^2 \equiv 1 \pmod 5$. Since q is a prime-power $\equiv 1 \pmod 3$, and since $|\Gamma_R|$ satisfies (*), $\Gamma_R \simeq A_4 \Rightarrow q = 7, 13$, or 19, $\Gamma_R \simeq S_4 \Rightarrow q = 7, 25$, or 31, and $\Gamma_R \simeq A_5 \Rightarrow q = 19, 25, 31, 49, 61, 79$, or 109. For (i) to (iv), $|R^\Gamma| = q(q - 1)$, $q(q - 1)/2$, $q(q + 1)/2$, $q(q - 1)$ respectively. Thus since $|\bar{g}| = q(q - 1)$, (iii) can only occur if the remaining orbits on $\bar{g}$ are of types (v) to (vii). Unkelbach now shows that (v), (vi), (vii) cannot occur, for the most part by a straightforward analysis of the orbit lengths. (For example, if $\Gamma_R \simeq A_5$ with $q = 25$, then $|\bar{g}| = 600$, $|R^\Gamma| = 130$, and possible lengths for the remaining orbits on $\bar{g}$ are 300 (from (ii)), 325 (from (iii) and (vi)), and 130 (from (vii)). However, the equation $130x + 300y + 325z = 600$ has no solution in non-negative integers with $x \geq 1$.) The case $\Gamma_R \simeq A_5$, $q = 19$, requires a special argument, involving the Sylow 5-groups of A_5. We are finally left with (i), (ii), and (iv). Since Σ_R contains the involution σ, Σ_R is the complete inverse image of Γ_R in Σ, and we obtain the groups listed in (2).

(3) $Z(\Delta_{Q,g})$ contains a subgroup θ of index 2 which consists of (Q, g) homologies.

Let $R \in \bar{g}$. Then $N_\Sigma(\Sigma_R)$ is transitive on the fixed points of Σ_R in the orbit R^Σ. In cases (a) and (c) of (2), Σ is transitive on $\bar{g}$ and it is known that $|N_\Sigma(\Sigma_R)| = 2|\Sigma_R|$; thus Σ_R has exactly 2 fixed points. In case (b), Σ has 2 orbits and $|N_\Sigma(\Sigma_R)| = |\Sigma_R|$, so that Σ_R has exactly one fixed point in each orbit. Let R, R' be the fixed points of Σ_R.

Now $\Sigma_R \subseteq \Delta_{Q,g}$, so that $Z(\Delta_{Q,g})$ commutes elementwise with Σ_R. Thus, at most, $Z(\Delta_{Q,g})$ exchanges R and R'. But $Z(\Delta_{Q,g})$ induces (faithfully) a group of (Q, g) homologies of the desarguesian plane π_0, and hence is cyclic. Thus $Z(\Delta_{Q,g})$ contains a cyclic subgroup θ of index 2 which fixes R, R' and hence fixes all points in R^Γ, R'^Γ. This means that θ consists of (Q, g) homologies.

(4) $\Delta_{P,\ell}$ contains a cyclic subgroup Ω of index 4.

Consider $\Omega_1 = \theta Z$, with θ, of order $(q - 1)/6$, as in (3), and Z the cyclic subgroup of Σ_R, with order $(q + 1)$ or $(q + 1)/2$, whose existence is assured by (2). Since $\theta \subseteq Z(\Delta_{Q,g})$, Ω_1 is abelian, and since $|\theta \cap Z| \leq 2$ (because $(q - 1, \ q + 1) = 2$), $|\Omega_1| = |\Delta_{P,\ell}|/d$, where d = 2 or 4. Since $\Delta_{P,\ell}$ is a Frobenius complement and Ω_1 is abelian, Ω_1 is cyclic by (2). Hence Ω_1 contains a cyclic subgroup Ω of order $|\Delta_{P,\ell}|/4$.

At this point we have a good deal of information about $\Delta_{P,\ell}$. In order to pin it down completely, Unkelbach considers "Singer cycles" of $\Delta_{Q,g}$. The group G = GL(2, q) contains a conjugacy class of cyclic subgroups T_1 of order $q^2 - 1$ which are sharply transitive on the non-zero vectors of the 2-dimensional vector-space over GF(q). Hence the group $\Delta_{Q,g} \simeq GL(2, q)/Z_0$ contains cyclic subgroups T of order $(q^2 - 1)/3$, the "Singer cycles" of $\Delta_{Q,g}$. These have the following properties, which can be deduced from the corresponding properties of the groups T_1 (see [19], pp.

187-188). Denote the normalizer in $\Delta_{Q,g}$ of T by N(T). Then (i) N(T) = T<i> (semidirect), where i is an involution not in $\Sigma = SL(2, q)$; (ii) $|T \cap \Sigma| = q + 1$; (iii) $|N(T) \cap \Sigma| = 2(q + 1)$; (iv) if $\tau_1 \in \Delta_{Q,g}$ and has odd order dividing q + 1, it determines a Singer cycle T such that $\tau_1 \in T$; (v) if T is a Singer cycle, if $\tau \in T$, and if the order of τ does not divide q - 1, then $\langle\tau\rangle$ and T have the same normalizer in $\Delta_{Q,g}$.

(5) $\Delta_{P,\ell} \subseteq N(T)$, where T is a Singer cycle of $\Delta_{Q,g}$.

Consider $\Delta_{P,\ell}$ as a permutation group on the 4 cosets of the cyclic subgroup Ω defined in (4). Let K be the kernel of the representation, so that K is the intersection of all conjugates of Ω in $\Delta_{P,\ell}$. Then $|\Delta_{P,\ell}/K|$ divides 4! = 24. If $|K|$ divides q - 1, so that $|K|x = q - 1$, then x divides $3 \times 24/q + 1$, since $|\Delta_{P,\ell}| = (q^2 - 1)/3$. This can only happen if q = 7 (since we assume q is a prime-power $\equiv 1 \pmod 3$)). For q = 7, $|\Delta_{P,\ell}| = 16$ and $\Delta_{P,\ell}$ is cyclic or a generalized quaternion by (1). The Sylow 2-groups of N(T), where T is a Singer cycle, are in this case Sylow 2-groups of $\Delta_{Q,g}$ (both have order 32). Hence, by the Sylow theorems, $\Delta_{P,\ell} \subseteq N(T)$ for some T, if q = 7. If $q \neq 7$, then by properties (iv) and (v) above, there exists $\tau \in K$ and a Singer cycle T of $\Delta_{Q,g}$ such that $\langle\tau\rangle$ and T have the same normalizer N(T) in $\Delta_{Q,g}$. But K is normal in $\Delta_{P,\ell}$ and $\langle\tau\rangle$ is characteristic in K since K is cyclic. Hence $\Delta_{P,\ell} \subseteq$ N(T).

If $q \equiv 3 \pmod 4$, the Sylow 2-groups S of N(T) are quasidihedral (i.e. are of the form $\langle A, i\rangle$ with $A^{2^n} = i^2 = 1$ and $iAi = A^{-1+2^{n-1}}$, where here $\langle A\rangle$ is a Sylow 2-group of T). They contain exactly one cyclic subgroup of index 2, namely $\langle A\rangle$, and one generalized quaternion group U of index 2, namely $\langle A^2, iA\rangle$. It follows that N(T) contains, to within conjugacy, just 2 subgroups of index 2 which have a cyclic or generalized quaternion Sylow 2-group, namely T and $O(T)\cdot U$, where O(T) denotes the maximal normal odd-order subgroup of T. Thus $\Delta_{P,\ell} = T$ or $O(T)\cdot U$.

If $q \equiv 1 \pmod 4$, the Sylow 2-groups S of N(T) are of the form $\langle A, i\rangle$ with $A^{2^n} = i^2 = 1$ and $iAi = A^{1+2^{n-1}}$, where again $\langle A\rangle$ is a Sylow 2-group of T. Besides $\langle A\rangle$, they contain exactly one cyclic subgroup $W = \langle iA\rangle$ of index 2, and no generalized quaternion subgroup of index 2. Hence here $\Delta_{P,\ell} = T$ or $O(T)\cdot W$ (a non-abelian group).

Thus finally, $\Delta_{P,\ell}$ is determined to within conjugacy in both cases, and, by (1), there are only 2 possibilities for π. Since the desarguesian planes and the Hughes planes satisfy the hypotheses of Theorem 1.9, they are the only planes which do so.

Theorem 1.2 follows from Theorems 1.8, 1.7, and 1.9.

REMARKS

1. Unkelbach states that cases (a), (b) of (2) correspond to the desarguesian and Hughes planes respectively,

and that (c) cannot occur. This is clear if $q \equiv 3 \pmod 4$. For if $\Delta_{P,\ell} \neq T$, the Sylow 2-groups of $\Delta_{P,\ell}$ are the Sylow 2-groups of $\Sigma \cap N(T)$, whence $|\Delta_{P,\ell} \cap \Sigma| = 2(q + 1)$, not $q + 1$. If $q \equiv 1 \pmod 4$, then case (c) does not occur. However, the fact that $2 \| |T \cap \Sigma|$ implies that the elements of order 4 in a Sylow 2-group of $\Sigma \cap N(T)$ are not in T, and hence must be $iA^{2^{n-2}}$ and its inverse: these are not in W. Hence $|\Delta_{P,\ell} \cap \Sigma| = q + 1$, not $2(q + 1)$, so that the Hughes planes must also arise from (a).

2. Theorem 1.2 could probably be proved without coordinates, using Unkelbach's approach. However, the assumption $q \equiv 1 \pmod 3$ is used quite strongly in the proof of Theorem 1.9, (2), and without it one would certainly have additional difficulties with those values of q for which there exists an irregular near-field of order q^2. (In particular, there are 4 non-isomorphic near-fields of order 11^2, including $GF(11^2)$.) Apart from $q = 7$, all the irregular near-fields have order q^2 with $q \not\equiv 1 \pmod 3$, and here these are handled by Dembowski's proof, in which it is unnecessary to distinguish between non-isomorphic planes of the same order.

ADDENDUM. The special Hughes plane of order 7^2 has the interesting property that the group generated by all its elations is isomorphic to SL(3, 7), not PSL(3, 7) (cf. Theorem 3.2(d)). The following result, which characterizes the plane in question, was proved by Lüneburg during the course of this conference.

THEOREM (Lüneburg). Let π be a plane of order q^2, where $q \equiv 1 \pmod 3$, and let π_0 be a Baer subplane of π. Suppose that π admits a collineation group $\Delta \simeq SL(3, q)$ which fixes π_0 as a set. Then π is the special Hughes plane of order 7^2.

2. UNITALS AND PSU(3, q)

DEFINITION. U is a "unital of order s" <=> U is a configuration with exactly $(s^3 + 1)$ points, $s^2(s^2 - s + 1)$ lines, $(s + 1)$ points on each line, and s^2 lines on each point.

If U is embedded in a plane π of order s^2, then for each point $X \in U$ there is a unique line T(X) on X, the "tangent line at X," such that $T(X) \not\in U$. Every line in π is either a tangent or a line of U, since $s^2(s^2 - s + 1) + (s^3 + 1) = s^4 + s^2 + 1$.

The following are examples of unitals. (1) Let θ be a unitary polarity of the desarguesian plane of order q^2. Then the absolute points and non-absolute lines of θ form a unital of order q. (The point P is "absolute" <=> $P \in \theta(P)$, and dually.) (2) For each Ree group G(q), $q = 3^{2r+1}$, there is a unital of order q which admits G(q) as a 2-transitive automorphism group. However, such a unital cannot be embedded in a plane π of order q^2 in such a way that $G(g)$ extends to a collineation group of π (Lüneburg, 1966b). (3) Ganley [10] has shown that unitals occur in the planes over Dickson semifields. See also Seib [25].

DEFINITION. Let θ be a unitary polarity of the desarguesian plane of order q^2. Then $PSU(3, q)$ is the group of all collineations in $PSL(3, q^2)$ which commute with θ.

THEOREM 2.1. Let θ be a unitary polarity of the desarguesian plane of order q^2, and write $\Gamma = PSU(3, q)$, where Γ commutes with θ. Then: (i) The absolute points and non-absolute lines of θ form a unital, U, of order q. (ii) $|\Gamma| = (q^3 + 1)q^3(q^2 - 1)/d$, where $d = (q + 1, 3)$. (iii) Γ is 2-transitive on the points of U. (iv) If $X \in U$, Γ_X contains a normal subgroup Q of order q^3 which is sharply transitive on $U - \{X\}$; the centre of Q is an elementary abelian group T of order q, and the elements of T are $(X, T(X))$ elations. (v) If $\ell \in U$, $|\Gamma_\ell| = (q + 1)^2 q(q - 1)/d$ and Γ_ℓ contains a normal cyclic subgroup H of order $(q + 1)/d$ whose elements are $(\theta(\ell), \ell)$ homologies. (vi) If q is even, each involution in Γ has centre X and axis $T(X)$ for some $X \in U$. If q is odd and if σ is an involution in Γ, then σ has axis ℓ and centre $\theta(\ell)$ for some $\ell \in U$; $C_\Gamma(\sigma)$ has 2 orbits on U, of lengths $(q + 1)$ and $(q^3 - q)$ respectively (namely $U \cap \ell$ and $U - (U \cap \ell)$). (vii) Γ is simple if $q > 2$.

For proofs see Huppert [19], pp.242-244, or Hughes and Piper [18], pp.57-63. Most of the results can be derived quite easily from the fact that, in homogeneous coordinates, $\theta(x_1, x_2, x_3) = [x_3^q, x_2^q, x_1^q]$ is a canonical form for θ.

A polarity of a desarguesian plane is "unitary" $\Leftrightarrow$ it corresponds to a hermitian form in the associated 3-dimensional vector-space. For a general plane of order s^2 (not necessarily desarguesian), a polarity is said to be "unitary" $\Leftrightarrow$ its absolute points and non-absolute lines form a unital of order s. The two definitions are equivalent for the desarguesian planes of order q^2. By a theorem of Baer, 1946a, a plane of order n can admit a unitary polarity only if n is a square.

The aim of this section is to describe briefly the proofs of two theorems concerning unitals and PSU(3, q). The first is of type I and the second of type II, in the sense of the Introduction.

THEOREM 2.2 (Hoffer [17]). *Suppose that* π *is of order* q^2 *and admits a collineation group* $\Delta \simeq$ PSU(3, q). *Then* (i) π *is desarguesian*, (ii) *for some unitary polarity* θ *of* π, Δ *acts on* π *as* Γ *acts in Theorem* 2.1.

THEOREM 2.3 (Kantor [21]). *Let* π *be of order* q^2, θ *a unitary polarity of* π, *and* U *the unital formed by the absolute points and non-absolute lines of* θ. *Suppose that* π *admits a collineation group* Δ *which commutes with* θ *and is flag-transitive on* U. *Then* π *is desarguesian and* $\Delta \supseteq PSU(3, q)$.

We first state a theorem due to Seib, which plays an important role in the proofs of Theorems 2.2 and 2.3. It is proved by a simple but ingenious counting argument.

THEOREM 2.4 (Seib [25]). *Let* U *be a unital of order* s, *embedded in a plane* π *of order* s^2. *Let* σ *be a Baer involution of* π *which fixes* U *as a set*. *Then* (i) U *contains exactly* (s + 1) *fixed points of* σ, (ii) *these points are collinear if* s *is even but form an oval if* s *is odd*.

PROOF OUTLINE FOR THEOREM 2.2. Assume $q > 2$ (since otherwise π is certainly desarguesian).

(1) Δ fixes no point or line and has a point-orbit O of length $q^3 + 1$. Δ is 2-transitive on O, and acts on O as $\Gamma = PSU(3, q)$ acts on the subgroups Γ_X, $X \in U$, in the representation given in Theorem 2.1.

Since $q > 2$, Δ is simple, and hence is faithful on each point or line orbit of length > 1. If Δ fixes a line ℓ, then it fixes all points on ℓ, since it is known that Δ cannot be represented non-trivially on less than $q^3 + 1$ points if $q \neq 5$, and on less than 50 points if $q = 5$. Similarly Δ must fix each line on a point $X \in \ell$, so that Δ fixes π pointwise. Thus Δ has no fixed line, and dually no fixed point. The maximal subgroups of PSU(3, q) were determined by Mitchell, 1911, and Hartley, 1926, so that all possible orbit lengths x with $1 < x \le q^4 + q^2 + 1$ can be listed. By analysing the combinations of these lengths (cf. the proof of Theorem 1.9, (2)), Hoffer shows that there must exist a point-orbit O of length $q^3 + 1$. The other statements in (1) now follow from the fact that PSU(3, q) has only one transitive representation of degree $q^3 + 1$.

(2) Let U consist of the points of O together with those lines which contain at least 2 points of O. Then U is a unital of order q, and Δ fixes U.

This follows easily from the fact that Δ is 2-transitive on the points of U.

(3) The involutions in Δ are perspectivities.

Suppose that σ is a Baer involution in Δ, with fixed subplane π'. By Theorem 2.4, σ fixes exactly $q + 1$ points of U. But if q is even, by (1) and Theorem 2.1(vi), σ fixes only one point of U. Hence q is odd, and by Theorem 2.4, $U \cap \pi'$ is an oval. But, by (1), $C_\Delta(\sigma)$ must be the stabilizer of the set of points $\ell \cap U$ for some $\ell \in U$. Since $C_\Delta(\sigma)$ has 2 orbits of length $q + 1$ and $q^3 - q$, by Theorem 2.1(vi), and since $C_\Delta(\sigma)$ must fix the set of $q + 1$ points in $U \cap \pi'$, ℓ must contain $U \cap \pi'$, a contradiction. The result now follows from Baer's theorem.

(4) π is isomorphic to an incidence structure $\pi(\Delta)$ whose points, lines, and incidences are determined within the group Δ.

For $q = 2^r$, the groups T of Theorem 2.1(iv) are elementary abelian 2-groups, and hence, by (3), consist of commuting involutory elations. It is easily seen that each such T is the group of all (X, T(X)) elations in Δ for some $X \in U$, and that each $X \in U$ corresponds to a unique group T. Hoffer shows that the map $X \rightarrow T(X)$ can be extended to a unitary polarity θ of π, with associated unital U. He then proves that each group H in Theorem 2.1(v) is the group of all $(\theta(\ell), \ell)$ homologies in Δ for some $\ell \in U$, and that each $\ell \in U$ corresponds to a unique group H. Thus each point $X \in U$ (line T(X)) can be associated with the corresponding group T, and each line $\ell \in U$ (point $\theta(\ell)$), with the corresponding group H. Furthermore, it is shown that a given point and line are incident if and only if the associated subgroups stand in a prescribed relation (for example, the point $X \in U$ is on the line $\ell \in U \Longleftrightarrow$ the corresponding groups T, H centralize each other).

For q odd, the involutions σ in Δ are homologies by (3), and in fact are (P, ℓ) homologies with $\ell \in U$, $P \notin U$, where P is uniquely determined by ℓ. The map $\ell \rightarrow P$ can be extended to a unitary polarity θ of π, with associated unital U. Once again, each point $X \in U$ (line T(X)) corresponds to a unique group T consisting of all (X, T(X)) elations in Δ; each line $\ell \in U$ (point $\theta(\ell)$) now corresponds to the unique involutory $(\theta(\ell), \ell)$ homology. Incidence can be described by group-theoretic relations between the groups T and $\langle\sigma\rangle$ (for example, $X \in U$ is on $\ell \in U \iff \sigma$ normalizes T).

Let $\bar{\pi}$ be the desarguesian plane of order q^2, and let $\bar{\theta}$ be a unitary polarity of $\bar{\pi}$. Let $\bar{\Gamma} \simeq PSU(3, q)$ be the subgroup of $PSL(3, q^2)$ which commutes with $\bar{\theta}$. Then $\bar{\pi}$ can be represented as $\bar{\pi}(\bar{\Gamma})$. Hoffer shows that $\pi(\Delta) \simeq \bar{\pi}(\bar{\Gamma})$, whence it follows that π is desarguesian and that Δ acts on π as $\bar{\Gamma}$ acts on $\bar{\pi}$.

We shall describe the proof of Theorem 2.3 only for even q: the proof for q odd is much more difficult, and uses a great deal of group theory. The proof for q even gives at least some idea of the type of arguments used.

PROOF OUTLINE FOR THEOREM 2.3, q EVEN

(1) Δ is faithful on U and primitive on the points of U.

If $\delta \in \Delta$ fixes all points of U, it must also fix all lines on each point of U, whence $\delta = 1$. Since Δ is flag-transitive on U by hypothesis, it is primitive on the points of U by a theorem of Higman and McLaughlin, 1961.

(2) Let $O(\Delta)$ be the maximal normal subgroup of odd order in Δ. If $O(\Delta) \neq \{1\}$, then $q = 2$.

By the Feit-Thompson theorem, 1963, $O(\Delta)$ is solvable. Suppose $O(\Delta) \neq \{1\}$, and consider the last non-trivial subgroup S in the commutator series of $O(\Delta)$. For some prime p, S contains an elementary abelian p-group E which is characteristic in S, hence characteristic in $O(\Delta)$, hence normal in Δ. By (1), E is sharply transitive on the $q^3 + 1$ points of U. But $p^r = q^3 + 1 \Rightarrow p \mid (q + 1, q^2 - q + 1) = 3$. It follows easily that $p^r = 3^2$, whence $q = 2$.

Henceforth we assume $q > 2$, since otherwise π is desarguesian.

(3) Δ contains an involutory $(X, T(X))$ elation, for each $X \in U$.

Let Σ be a Sylow 2-group of Δ. Then, since $|U|$ is odd, Σ fixes some point $X \in U$. If all involutions in Δ are Baer involutions, then there exists a Baer involution $\sigma \in Z\Sigma$. By Theorem 2.4, the $q + 1$ fixed points of σ in U are on a line $\ell \in U$, where $X \in \ell$. Clearly Σ fixes ℓ. But Δ_X is transitive on the q^2 lines of U on X, so that $|\Delta_X| = q^2 |\Delta_{X,\ell}|$. This implies that Σ is not of maximal 2-power order in Δ_X, since q is even. Hence, by Baer's theorem on involutions, Δ contains involutory elations. It is easily verified that an involutory elation which fixes $X \in U$ must have centre X and axis T(X). Since Δ is transitive on the points of U, it contains an involutory $(X, T(X))$ elation for each $X \in U$.

(4) If Δ contains a Baer involution fixing $X \in U$, then all involutory $(X, T(X))$ elations commute.

Let σ be a Baer involution which fixes $X \in U$, and hence fixes the points of $\ell \cap U$ for some ℓ on X, $\ell \neq T(X)$, by Theorem 2.4. Let α, β be involutory $(X, T(X))$ elations.

Then $(\sigma\alpha)^2$ is an $(X, T(x))$ elation which fixes all points of $\ell \cap U$, and hence is the identity. Thus $\sigma\alpha = \alpha\sigma$, and clearly $\sigma\alpha$ is a Baer involution. Similarly, $\alpha\beta = \beta\sigma$ and $(\sigma\alpha)\beta = \beta(\sigma\alpha)$. Thus $(\alpha\beta)\sigma = \sigma(\alpha\beta) = (\sigma\alpha)\beta = \beta(\sigma\alpha) = (\beta\alpha)\sigma$, whence $\alpha\beta = \beta\alpha$.

(5) Δ contains a normal subgroup N, transitive on the points of U, all of whose involutions are elations.

By (3), Δ contains an involutory $(X, T(X))$ elation for each $X \in U$. If Δ also contains Baer involutions, then by (4), for given $X \in U$ the involutory $(X, T(X))$ elations form, together with the identity, an elementary abelian 2-group which is clearly normal in each Sylow 2-group of Δ_X. The existence of N now follows from a (deep group-theoretic) theorem of Shult [24] (see also Goldschmidt [11]).

(6) Δ contains a subgroup $K \simeq PSL(2, q^3)$, $Sz(q^{3/2})$, or $PSU(3, q)$, acting on the points of U in the usual 2-transitive representation; q is a power of 2. (Here Sz(s) denotes the Suzuki group of order $(s^2 + 1)s^2(s - 1)$, where $s = 2^{2e+1}$.)

Each involution in N is in fact an $(X, T(X))$ elation for some $X \in U$, as is easily verified. Also $O(N) = \{1\}$, by (2), since otherwise $O(\Delta) \neq \{1\}$, and we are assuming $q > 2$. Since each involution in N fixes exactly one point of U, it now follows from a theorem of Bender [2] that N contains a normal subgroup K isomorphic to one of the groups listed in (6), and that K acts on the points of U in the usual 2-transitive representation. If $K \simeq PSL(2, q^3)$ or $Sz(q^{3/2})$, then, for $X \in U$, Δ_X contains a group of order q^3 or $q^{3/2}$ respectively which is an elementary abelian 2-group and hence consists of $(X, T(X))$ elations. This is impossible, since Δ contains at most q $(X, T(X))$ elations. Hence $K \simeq PSU(3, q)$, and we can apply Theorem 2.2.

REMARK. The proof of Theorem 2.2 relies on the comprehensive analysis of the maximal subgroups of PSU(3, q) made by Mitchell, 1911, and Hartley, 1926, but requires no modern group theory; the proof of Theorem 2.3 involves three recent and very deep group-theoretic results. This contrast is typical of the difference between the group-theoretic methods used in solving type I and type II problems.

3. COLLINEATION GROUPS WHICH CONTAIN ELATIONS

Finite collineation groups which contain non-trivial elations have been studied extensively, in particular by Wagner, 1959, and Piper, 1963, 1965, 1966a. An account of this fundamental work is given in [5], §4.3. Here we shall describe the main results of a recent paper by Hering [16].

THEOREM 3.1. *Let π be a finite plane and Δ a collineation group of π. Let Σ be the group generated by all elations in Δ. Let $\underline{A} = \{a \mid a$ is the axis of a non-trivial elation in $\Delta\}$, $\underline{Z} = \{Z \mid Z$ is the centre of a non-trivial elation in $\Delta\}$. Then, up to duality, one of the following holds:*

I $\underline{Z} = \underline{A} = \phi$.

II $\underline{Z} = \{Z\}$, $\underline{A} = \{a\}$, *for some* Z, a *with* $Z \in a$.

III *For some* P, ℓ *with* $P \notin \ell$, $\underline{Z} \subseteq \ell$ *and* $\underline{A} = \{XP \mid X \in \underline{Z}\}$.

IV *For some* a, $\underline{A} = \{a\}$, $\underline{Z} \subseteq a$ *and* $|\underline{Z}| > 1$.

V *For some* Z, a *with* $Z \in a$, $\underline{Z} \subseteq a$, $\underline{A} \subseteq [Z]$ (*the set of lines on* Z), $|\underline{Z}| > 1$ *and* $|\underline{A}| > 1$.

VI *For some* a, $\underline{Z} \subseteq a$ *but* $\underline{A} \not\subseteq [Z]$ *for any* $Z \in \underline{Z}$.

VII *The points of* $\underline{Z}$ *and the lines of* $\underline{A}$ *form a subplane of* π.

VIII *There exists a subplane* π' *of order* 4 *in* π *and a hyperoval* H *of* π' (*i.e. a set of* 6 *points, no* 3 *collinear*), *such that* $\underline{Z} = \pi' - H$ *and* $\underline{A}$ *is the set of all lines which meet* H *in* 2 *distinct points*.

IX *If* $x \in \underline{A}$, *then the stabilizer* Δ_x *is of type II*. Σ *does not fix any point-line pair*.

X *If* $x \in \underline{A}$, *then* Δ_x *is of type II, IV or IVd (the dual of IV)*. *There exists* $x \in \underline{A}$ *such that* Δ_x *is of type IV*. *Also* Σ *fixes no centre or axis*.

REMARKS

1. The theorem is a generalization of the "Lenz classification" ([15], p.126). Note that types I to VII are entirely analogous to the Lenz classes with the corresponding number.

2. The theorem is stated in terms of the sets $\underline{Z}$, $\underline{A}$. However, it can also be regarded as a classification according to the figure formed by those incident point-line

pairs (Z, a) such that Δ contains a non-trivial (Z, a) elation. For if $Z \in \underline{Z}$ and $a \in \underline{A}$, there exist a', Z' such that Δ contains a non-trivial (Z, a') elation α and a non-trivial (Z', a) elation β. If $Z \in a$, and if $Z \neq Z'$, $a \neq a'$, then it is easily verified that $\alpha\beta\alpha^{-1}\beta^{-1}$ is a non-trivial (Z, a) elation (Wagner, 1959). Thus Δ contains a non-trivial (Z, a) elation <=> $Z \in \underline{Z}$, $a \in \underline{A}$, and $Z \in a$.

PROOF OF THEOREM 3.1

(1) If $\underline{Z} = \phi$ and if Σ fixes $a \in A$, then Δ is of type II, IV, IVd, V, or VI.

If Σ fixes a, then $\underline{Z} \subseteq a$. (If $\delta \in \Delta - \Sigma$ does not fix a, then Σ fixes δ(a), since $\Sigma \triangleleft \Delta$. Thus Σ fixes $a \cap \delta(a)$, which is thus the only possible centre.) Δ is of type II or IV if $\underline{A} = \{a\}$, of type II or IVd if $|\underline{Z}| = 1$, and is otherwise of type V or VI.

We can now assume that Σ fixes no axis, and dually no centre.

(2) For each $a \in A$, Δ_a is of type II, IV, IVd, or VI. This follows immediately from (1).

(3) If Σ fixes no centre or axis, and if there exists $a \in \underline{A}$ such that Δ_a is of type V or VI, then Δ is of type VII or VIII.

If Σ fixes a point F, then $F \notin \underline{Z}$ by hypothesis, so that $F \in a$ for each $a \in \underline{A}$. But this implies $F \in \underline{Z}$, since by hypothesis $|\underline{A}| \geq 2$. Hence Δ fixes no point or line.

The conclusion now follows from results of Piper, 1965, 1966a. (In 1966a, Piper does not exclude the possibility that the centres and axis of elations in Δ should form a set of disjoint subplanes of order 2. Hering states that this can be ruled out by applying a theorem of Bender [2].)

We can now assume:

(*) $\underline{Z} \neq \phi$. If $a \in \underline{A}$ then Δ_a is of type II, IV, or IVd.

Note that (*) => (*)'.

(*)' If $Z \in \underline{Z}$ and if $a \in \underline{A}$, then Δ_Z, Δ_a are of types II, IV, or IVd.

(4) If Σ fixes no centre or axis and if Δ satisfies (*), then Δ is of type III, IX, X, or Xd.

Assume Δ is not of type X or Xd. Then for each $a \in \underline{A}$, Δ_a is of type II. If Σ fixes (P, ℓ), then $P \notin \ell$, for otherwise it is easily shown that $P \in \underline{Z}$ or $\ell \in \underline{A}$. If Σ fixes some (P, ℓ) with $P \notin \ell$, then Δ is of type III, and otherwise Δ is of type IX.

REMARK. Examples of groups of each of the types I to X (and, of course, their duals) can be found in desarguesian planes. In particular, the plane of order 4 admits a group of type VIII (Piper 1966a), in the desarguesian plane of order q^2, PSU(3, q) is of type IX (see section 2), and in the desarguesian plane of order $q = 2^r$, the group generated by the elations which fix a given conic ($\simeq$ PSL(2, q)) is of type X (see Dembowski [5], p.185).

The remainder of [16] is concerned with the structure of the group Σ.

THEOREM 3.2. *Let* π *be a finite plane,* Δ *a collineation group of* π, *and let* Σ *be the group generated by all elations in* Δ. *Let* p *be the smallest prime such that* Δ *contains an elation of order* p. *Let* $K = \{\delta \in \Delta \mid \delta$ *fixes* $\underline{Z}$ *pointwise*$\}$. *Then*: (a) *If* Δ *is of type IV,* Σ *is an elementary abelian p-group*. (b) *If* Δ *is of type V,* Σ *is a p-group of class* 2. (c) *If* Δ *is of type VI and if* T *is the group of all elations with axis* a, *then either* (i) $\Sigma/T \simeq SL(2, q)$, *where* $q = p^r$, (ii) $p = 3$ *and* $\Sigma/T \simeq SL(2, 5)$, (iii) $p = 2$ *and* $\Sigma/T \simeq Sz(2^{2r+1})$, *or* (iv) $p = 2$ *and* 4 *does not divide* $|\Sigma/T|$. (d) *If* Δ *is of type VII, then* $\Sigma/\Sigma \cap K \simeq PSL(3, q)$, *where* $q = p^r$, *and every non-trivial elation in* Δ *has order* p. *Also* $\Sigma \cap K \subseteq \Sigma' \cap Z(\Sigma)$ (*where* Σ' *is the commutator group of* Σ). (e) *If* Δ *is of type VIII, then* $\Sigma/\Sigma \cap K \simeq A_6$, *and every non-trivial elation in* Δ *has order* 2. *Again* $\Sigma \cap K \subseteq \Sigma' \cap Z(\Sigma)$.

For proofs, see (a) Baer, 1942, (b) Hering, 1963, (c) Hering [15], (d) Piper, 1965 and 1966a, (e) Piper 1966a. The fact that $\Sigma \cap K \subseteq \Sigma' \cap Z(\Sigma)$ for types VII and VIII is proved in [16]. It means that Σ is a "Schur extension" of $\Sigma^* = \Sigma/\Sigma \cap K$, so that the theory of the Schur multiplier can be used in determining Σ from Σ^*. (See, for example, Huppert, [19], p.628.)

The corresponding problem for classes III, IX, X is not completely solved (and appears to be difficult). However, if Δ contains involutory elations, Hering is able to determine the group generated by all involutory elations in Δ. The proof depends on the following deep group-theoretic result, which is also due to Hering [13].

THEOREM 3.3. *Let Γ be a finite group which is transitive on the set* W, *where* $|W| > 1$. *Suppose that for* $X \in W$, Γ_X *contains a normal subgroup* N *of even order such that* $N_Y = \{1\}$ *for all* $Y \in W - \{X\}$. *If* Φ *is the normal closure of* N *in* Γ (i.e. *the group* $\langle N^\gamma \mid \gamma \in \Gamma \rangle$), *then either* $\Phi = N \cdot O(\Phi)$ *and* N *is a Frobenius complement*, *or* $\Phi \simeq SL(2, q)$, $Sz(q)$, $SU(3, q)$, *or* $PSU(3, q)$, *where* $q = 2^r \geq 4$.

THEOREM 3.4. *Suppose* Δ *satisfies the condition* (*), *fixes no centre or axis, and contains involutory elations. Let* $\bar{\Sigma}$ *be the group generated by the involutory elations in* Δ. *If 4 divides* $|\bar{\Sigma}|$, *then* $\bar{\Sigma} \simeq SL(2, q)$, $Sz(q)$, $SU(3, q)$, *or* $PSU(3, q)$, *where* $q = 2^r \geq 4$.

PROOF. Let α be an involutory (X, a) elation in Δ. By (*), we can assume

(**) for each axis a' on X, X is the only centre on a'. (Otherwise the dual holds, and the proof is dualized throughout.)

Let $W = \{\delta(x) \mid \delta \in \Delta\}$. Then if $Y \in W$, Y also satisfies (**). Let N be the group of all elations in Δ which have centre X. Then $N \triangleleft \Delta_X$, $|N|$ is even, and if $Y \in W - \{X\}$, $N_Y = \{1\}$ by (**). Thus $|W|$ is odd, and $|W| > 1$ since X is not fixed by Δ. If β is any involutory elation in Δ, β must fix some $Z \in W$. It follows from (**), applied to Z, that Z is the centre of β. Let Φ be the normal closure of N in Δ. Then $\bar{\Sigma} \subseteq \Phi$. We can now apply Theorem 3.3 to Δ, W, N, and Φ. If $\Phi = N \cdot O(\Phi)$ with N a Frobenius complement, then N contains a unique subgroup N_1 of order 2, and all involutions in Φ belong to $N_1 \cdot O(\Phi)$, whence 4 does not divide $|\bar{\Sigma}|$. Otherwise, $\Phi \simeq SL(2, q)$, $Sz(q)$, $SU(3, q)$, or $PSU(3, q)$, where $q = 2^r \geq 4$, and finally $\bar{\Sigma} = \Phi$ since these groups contain no proper normal subgroup of even order.

REMARKS

1. If Δ contains involutory elations and is not of type II, then $\bar{\Sigma}$ is determined, by Theorems 3.2 and 3.4. This is a very important and impressive result, which obviously has many possible applications to problems concerning collineation groups of finite planes of even order.

2. One cannot reasonably expect to obtain an analogous result for type II by these methods. For the known finite planes, groups of type II are elementary abelian.

3. Another application of Theorem 3.3 will be given in section 4, and it seems that it may well be useful in other geometrical contexts.

4. COLLINEATION GROUPS WHICH CONTAIN HOMOLOGIES

Homologies are in general more difficult to deal with than elations, and the results to be discussed here are not yet as complete as those described in section 3. We shall mainly be concerned with groups which contain involutory homologies, but first mention two results which also apply to homology groups of odd order.

THEOREM 4.1 (Piper, 1967). *Let* π *be a finite plane, and let* Δ *be a collineation group which contains non-trivial homologies and fixes no point or line. If there exists a line* ℓ *such that* ℓ *contains the centre of a non-trivial homology in* Δ *and* ℓ *is the axis of 2 non-trivial homologies in* Δ *which have different centres, then the centres and axes of homologies in* Δ *form a desarguesian subplane* π_0, *and* Δ *restricted to* π_0 *contains the little projective group of* π_0.

THEOREM 4.2 (Brown [3], [4]). *Let* π *be a finite plane, and let* Δ *be a collineation group of* π *which fixes no point or line and contains homologies of order s. Then the number of orbits of centres (axes, centre-axis pairs) of order s homologies in* Δ *is* 1 *if* $s > 2$, *and is* 1 *or* 2 *if* $s = 2$.

Theorem 4.1 is deduced from Piper's results on elations (1965). The proof of Theorem 4.2 depends mainly on counting arguments; additional numerical information is given about the 2-orbit case for $s = 2$. Theorem 4.1 was used in the proof of Theorem 1.8.

REMARK. It is natural to ask whether Hering's "Lenz" classification of finite groups which contain elations can be refined to a "Lenz-Barlotti" classification of finite groups which contain elations and homologies (see Theorem 3.1 and [5], p.126). This should be feasible; the subdivision of type I would presumably include all the analogues of the Barlotti subclasses of Lenz class I, and perhaps others.

The following theorems are fundamental in the study of collineation groups which contain involutory homologies.

THEOREM 4.3 (see [5], p.120). Let α_1, α_2 be non-trivial (P_1, ℓ_1), (P_2, ℓ_2) homologies such that $\alpha_1\alpha_2 = \alpha_2\alpha_1$. Then either $P_1 = P_2$ and $\ell_1 = \ell_2$, or $P_1 \neq P_2$, $\ell_1 \neq \ell_2$ and $P_1 \in \ell_2$, $P_2 \in \ell_1$.

PROOF. If $\ell_1 = \ell_2$ but $P_1 \neq P_2$, $\alpha_1\alpha_2\alpha_1^{-1}\alpha_2^{-1}(P_2) \neq P_2$; and dually. If $P_1 \neq P_2$, $\ell_1 \neq \ell_2$, then $\alpha_1\alpha_2\alpha_1^{-1} = \alpha_2$ implies that α_1 fixes P_2, ℓ_2, whence $P_2 \in \ell_1$ and $\ell_2 \in P_1$.

THEOREM 4.4 (Ostrom, Lüneburg; see Dembowski [5], p.120). Let σ_1, σ_2 be involutory (P_1, ℓ_1) and (P_2, ℓ_2) homologies respectively, where $P_1 \neq P_2$, $\ell_1 \neq \ell_2$, $P_1 \in \ell_2$, and $P_2 \in \ell_1$. Then (a) $\sigma_3 = \sigma_1\sigma_2$ is an involutory (P_3, ℓ_3) homology, where $P_3 = \ell_1 \cap \ell_2$, $\ell_3 = P_1 \cup P_2$; (b) $\sigma_1\sigma_2 = \sigma_2\sigma_1$; (c) σ_i is the only involutory (P_i, ℓ_i) homology, $i = 1, 2, 3$.

The situation considered in the next theorem probably cannot occur: if it does, the theorem tells us a good deal about the group.

THEOREM 4.5 (Hering [14]). *Let* π *be a plane (not necessarily finite), and let* Δ *be a finite collineation group of* π. *Let* $\underline{Z} = \{Z \mid$ *there exists* $a \in Z$ *such that there is more than one involutory* (Z, a) *homology*$\}$. *Let* Σ *be the subgroup of* Δ *generated by the perspectivities with centre* $Z \in \underline{Z}$. *If* $|\underline{Z}| > 1$, *then* $\Sigma \simeq SL(2, q)$, $Sz(q)$, $SU(3, q)$, *or* $PSU(3, q)$, *where* $q = 2^r \geq 4$.

PROOF. (1) For each $Z \in \underline{Z}$, there exists exactly one line $a \not\ni Z$ such that π admits a non-trivial (Z, a) homology.

Assume that there exist 2 distinct involutory (Z, a) homologies α_1, α_2, and that α is a non-trivial $(Z, \bar{a})$ homology, with $\bar{a} \neq a$. Then $\langle \alpha_1, \alpha_2, \alpha_3 \rangle$ must act as a Frobenius group Γ on the lines of π not on Z, with Γ_a as a Frobenius complement. But then Γ_a must contain only one involution, a contradiction.

(2) Let $Z \in \underline{Z}$, and let N be the group of all (Z, a) homologies in Δ. Then $N \lhd \Delta_Z$, $|N|$ is even, and no element of N fixes $Z' \in \underline{Z}$ with $Z' \neq Z$.

The first two statements are obvious. If $\alpha \in N$ fixes $Z' \in \underline{Z}$, where $Z' \neq Z$ and $\alpha \neq 1$, then $Z' \in a$. Moreover, the unique homology axis a' of Z' must be on Z. But this contradicts Theorem 4.4(c). (The group N is in fact the group of all perspectivities in Δ with centre Z, since a non-trivial elation with centre Z would move the axis a.)

(3) The group Σ is transitive on $\underline{Z}$, and is the normal closure of N in Δ.

By (2), for each $Z \in \underline{Z}$, there exists an involution which fixes Z but no other point of $\underline{Z}$. Hence Σ is transitive on $\underline{Z}$ by a theorem of Gleason, 1956. It follows that Σ is the normal closure of N in Δ, i.e. the group generated by N and all its conjugates in Δ.

We can now apply Theorem 3.3. The case $\Sigma = N.O(\Sigma)$ does not occur, since N contains more than one involution. (Hering also shows that if Δ fixes no point or line, then $\Sigma \simeq SU(3, q)$ or $PSU(3, q)$.)

We shall conclude by giving a brief account of some recent results of Kantor [23]. The starting point is the following theorem.

THEOREM 4.6 (Kantor [23], [21]). *Suppose that the plane π (not necessarily finite) admits a finite collineation group Δ which contains commuting involutory homologies with distinct axes. Then (i) for each $(P, \ell) \in \pi$ with $P \notin \ell$, there exists at most one involutory (P, ℓ) homology in Δ, (ii) Δ contains no elementary abelian group of order 8 which is generated by 3 involutory homologies.*

PROOF. (i) Let σ_1, σ_2 be commuting (P_1, ℓ_1) and (P_2, ℓ_2)

involutory homologies respectively, where $\ell_1 \neq \ell_2$. Then, by Theorems 4.3, 4.4, $P_1 \neq P_2$, $\ell_1 \neq \ell_2$, $P_1 \in \ell_2$, $P_2 \in \ell_1$, and σ_i is the only involutory (P_i, ℓ_i) homology, i = 1, 2. Suppose that for some (P, ℓ), there exist 2 distinct involutory (P, ℓ) homologies τ_1, $\tau_2 \in \Delta$. Then $(P, \ell) \neq (P_i, \ell_i)$, i = 1, 2 and $\sigma_1\tau_1 \neq \tau_1\sigma_1$ by Theorems 4.3, 4.4. Consider $\langle\sigma_1, \tau_1\rangle$. If $\sigma_1\tau_1$ has odd order, σ_1 is conjugate to τ_1, which is not the case. If $\sigma_1\tau_1$ has even order 2n, then $(\sigma_1\tau_1)^n = \sigma$ is an involution which commutes with both σ_1 and τ_1; thus $\sigma \neq \sigma_1, \tau_1$. Moreover $\sigma' = \sigma_1\sigma$ is conjugate to τ_1, and hence is a homology. Hence, by Theorems 4.3 and 4.4, $\sigma = \sigma_1\sigma' = \sigma'\sigma_1$ is an involutory (Q, m) homology, where $Q \neq P_1$, $m \neq \ell_1$ because σ commutes with σ_1 but $\sigma \neq \sigma_1$. By Theorem 4.4 applied to σ and σ_1, σ is the only involutory (Q, m) homology. Hence $(Q, m) \neq (P, \ell)$, and by Theorems 4.3, 4.4 applied to σ and τ_1, τ_1 must be the only involutory (P, ℓ) homology, a contradiction.

(ii) Let K be a Klein 4-group generated by involutory (A_i, a_i) homologies $\alpha_i \in \Delta$, i = 1, 2. By (i), $A_1 \neq A_2$, $a_1 \neq a_2$, $A_1 \in a_2$, and $A_2 \in a_1$. Let $A_3 = a_1 \cap a_2$, $a_3 = A_1 \cup A_2$. Suppose that $\alpha \in \Delta$ is an involutory (A, a) homology which commutes with α_1, α_2 (hence with $\alpha_3 = \alpha_1\alpha_2$). Then α fixes A_i, a_i, i = 1, 2, 3. Hence $A = A_i$ and $a = a_i$ for some i = 1, 2, or 3. This implies $\alpha = \alpha_i$, i.e. $\alpha \in K$.

The importance of Theorem 4.6 lies in the fact that, with certain additional restrictions on Δ, it permits the application of the following deep characterization theorem.

THEOREM 4.7 (Alperin, Brauer, and Gorenstein [1]). *Let* Γ *be a finite simple (non-abelian) group which contains no elementary abelian 2-group of order 8*. *Then* $\Gamma \simeq PSL(2, q)$, $PSL(3, q)$, $PSU(3, q)$, M_{11}, A_7, *or* $PSU(3, 4)$, *where* q *is an odd prime power*. (Here M_{11} is the Matthieu group of degree 11, and A_7 the alternating group of degree 7.)

Before describing Kantor's main results, we mention that [23] contains a number of lemmas which are certainly of independent interest, one of which is the following.

THEOREM 4.7. *Let* Δ *be a collineation group of a finite plane of odd order*. *If* Δ *contains an involutory homology, then so does the centre of a Sylow 2-group of* Δ.

PROOF OUTLINE. Let σ be an involutory homology in Δ such that a Sylow 2-group Σ of $C_{\Delta}(\sigma)$ is as large as possible. Since Σ is not a Sylow 2-group of Δ, we can find $\Phi \subseteq \Delta$ such that $\Sigma \subseteq \Phi$ and Σ is of index 2 in Φ. We can construct an involution $\tau \in Z\Phi$ which is the product of commuting involutory homologies, and hence is a homology, by Theorems 4.3, 4.4. This contradicts the maximality of Σ.

The paper also contains useful results on groups which are generated by 2 involutory homologies σ, τ with $\sigma\tau$ of prime-power order,and on collineation groups of prime-power order which are inverted by an involutory homology. The following lemma will be used here in outlining the proof of Theorem 4.10(i). The proof is straightforward, and uses Theorem 4.4.

THEOREM 4.8. _Suppose that_ $\Gamma_1 \times \Gamma_2$ _is a collineation group, where_ Γ_1, Γ_2 _are dihedral groups of order_ $2p$ _with_ p _an odd prime, and suppose further that all involutions in_ Γ_1, Γ_2 _are homologies. Then all elements of_ Γ_1 _are perspectivities with the same centre or axis._

The main object of [23] is to study the following situation.

(*) Let π be a finite projective plane of odd order n, and let Δ be a collineation group of π which is generated by involutory homologies. Assume that Δ contains commuting involutory homologies with different axes, and that there is no involutory homology σ such that $O(\Delta) \in Z(\Delta/O(\Delta))$.

The assumption on $Z(\Delta/O(\Delta))$ excludes, for instance, the group of all projective collineations which fix a line in a desarguesian plane of odd order; it also plays an important role in the group-theoretic parts of the proofs.

We omit the statement of the first main theorem (Theorem A) of [23], since it is rather technical. The other main theorems (Theorems B and C) are stated in Theorems 4.9, 4.10.

THEOREM 4.9. *Let* π, Δ *satisfy* (*), *and suppose that* Δ *contains no Baer involutions. Then one of the following holds*: (a) $\Delta \simeq PSL(2, q)$, $PGL(2, q)$, $PSL\hat{}(2, 9)$, $PSL(3, q)$, $SL(3, q)$, $PSU(3, q)$, $SU(3, q)$, A_7, *or* $PSU(3, 4)$; *here* q *is an odd prime-power and* $PSL\hat{}(2, 9)$ *is a non-split central extension of* $PSL(2, 9)$ *by a group of order* 3. (b) Δ *has a 2-subgroup* $\langle\phi_1\rangle \times \langle\phi_2\rangle$ *with* $|\phi_1| = |\phi_2| \geq 2$ *such that* $N = O(\Delta)\langle\phi_1, \phi_2\rangle \triangleleft \Delta$ *and* $\Delta/N \simeq S_3$; Δ *induces* S_3 *on the set* $\{X, Y, Z\}$ *of centres of the involutions in the Klein group* Σ *of* $\langle\phi_1, \phi_2\rangle$ *and* $\Delta_{XYZ} = N \supseteq O(\Delta) \times \Sigma$.

The proof of Theorem 4.9 uses some very sophisticated group theory. The basic geometric distinction between cases (a) and (b) is that in (b) there exists a triangle whose vertex-set is fixed by Δ, while in (a) this is not the case. (In [22], §2, case (b) does not arise, since it is assumed there that Δ contains no proper normal subgroup of even order.) Essentially, the fact that π and Δ satisfy the hypotheses of Theorem 4.6 permits the application of Theorem 4.7 (though of course Δ is not assumed to be simple here). The case corresponding to $\Gamma \simeq M_{11}$ in Theorem 4.7 cannot occur here, and the groups which appear in Theorem 4.9(a) but not in Theorem 4.7 mostly arise as Schur extensions of the related groups in Theorem 4.7. The apparent "odd men out" in Theorem 4.9(a) are $PSU(3, 4)$ and A_7. While $PSU(3, 4)$ probably cannot occur, A_7 is in fact a subgroup of $PSU(3, 5)$.

THEOREM 4.10. _Let_ π, Δ _be as in Theorem 4.9_. (i) _If_ $\Delta \simeq PSL(3, q)$ _or_ $SL(3, q)$, _then_ π _contains a subplane_ π' _of order_ q _which is fixed by_ Δ _and on which_ Δ _induces_ $PSL(3, q)$. _All elations of_ π' _are induced by elations of_ π. _Also_ $q \mid n$, $(q - 1) \mid (n - 1)$, _and_ $(q + 1) \mid (n^2 - 1)$ (ii) _If_ $\Delta \simeq PGL(2, q)$ _with_ $(q, n) > 1$, _then_ π _contains a subplane_ π' _of order_ q _which has an orthogonal polarity preserved by_ Δ. _Moreover_ $q \mid n$, $(q - 1) \mid (n - 1)$, _and_ $(q + 1) \mid (n^2 - 1)$. (iii) _If_ $\Delta \simeq PSU(3, q)$ _or_ $SU(3, q)$ _with_ $(q, n) > 1$, _then_ π _contains a subplane_ π' _of order_ q^2 _which has a unitary polarity preserved by_ Δ. _Moreover_ $q \mid n^2$, $(q - 1) \mid (n - 1)$, _and_ $(q + 1) \mid (n^2 - 1)$. (iv) _If_ $\Delta \simeq A_7$ _and if_ $5 \mid n$, _then_ π _contains a subplane of order_ 5^2 _which is fixed by_ Δ _and on which_ Δ _induces_ A_7. (Statement (iv) is not contained in the preprint of [23], but was mentioned to me by Kantor in a letter.)

Theorem 4.10 (i), (iii) give very strong generalizations of some of the results described here in sections 1 and 2. (For the corresponding analogue of (ii), see Lüneburg, 1964c.) Note that in (i) there is no restriction whatever on the relation between n and q, and that the relations demanded in (ii) to (iv) are quite slight: essentially they ensure that, with $p^r = q$ in (ii), (iii) and $p = 5$ in (iv), a Sylow p-group of Δ fixes at least one point. It is, of course, assumed here that Δ contains no Baer involution.

The proofs of (i) to (iii) use the lemmas mentioned earlier, and do not involve deep group-theory. (Theorem 4.10 is a "Type I" theorem, in the sense of the Introduction.) We illustrate briefly by giving the proof of the first part of (i) for $\Delta \simeq PSL(3, q)$.

Let $q = p^r$. By considering PSL(3, q) acting on the desarguesian plane of order q, we can find a subgroup $\Gamma_1 \times \Gamma_2$ of Δ, where Γ_1, Γ_2 are dihedral of order 2p. By Theorem 4.8 we can assume without loss of generality that Γ_1 consists of perspectivities with centre X (otherwise dualize). Suppose Δ fixes X. Then since all involutions are conjugate in Δ and Δ is generated by its involutions, Δ consists of perspectivities with centre X: this contradicts (*). Let σ be an involution and τ an element of order p in Γ_1. Then $C_\Delta(\sigma) \subseteq \Delta_X$, and τ is centralized by some Sylow p-group Σ of Δ, whence $\Sigma \subseteq \Delta_X$. Since $C_\Delta(\sigma)$ is of index $q^2(q^2 + q + 1)$ in Δ, the group generated by $C_\Delta(\sigma)$ and Σ is of index at most $(q^2 + q + 1)$ in Δ. But since $\Delta \neq \Delta_X$, this in fact implies that Δ_X is of index precisely $q^2 + q + 1$ in Δ, and is isomorphic to the stabilizer in PSL(3, q) of a point in the desarguesian plane $\bar{\pi}$ of order q. Thus $|X^\Gamma| = q^2 + q + 1$, and Δ acts on X^Γ as does PSL(3, q) on the points of $\bar{\pi}$. The involution σ therefore fixes $q + 2$ points of X^Γ, and $C_\Delta(\sigma)$ permutes these in orbits of lengths 1 and q + 1. It follows that the axis of σ contains q + 1 points of X^Γ. Now Δ is 2-transitive on X^Γ, and it is easily deduced that the points of X^{Γ} and the lines which meet it in at least 2 points form a subplane π' of order q ([5], p.138). Finally, since Δ is faithful on X^{Γ} and $\Delta \simeq PSL(3, q)$, π' is desarguesian, by Dembowski, 1965c.

REFERENCES

1. J. Alperin, R. Brauer, and D. Gorenstein. Finite simple groups of 2-rank two (to appear).
2. H. Bender. Transitive Gruppen gerader Ordnung denen jede Involution genau einen Punktfestlasst. J. Algebra 17 (1971): 527-554.
3. J. Brown. Homologies in collineation groups of finite projective planes I. Math. Z. 124 (1972): 133-140.
4. —— Homologies in collineation groups of finite projective planes II. Math. Z. 125 (1972): 338-348.
5. P. Dembowski. Finite Geometries. Springer, New York (1968).
6. —— Zur Geometrie der Gruppen $PSL_3(q)$. Math. Z. 117 (1970): 125-134.
7. —— Generalized Hughes planes. Canad. J. Math. 23 (1971): 481-494.
8. —— Gruppen erhaltende quadratische Erweiterungen endlicher desarguesscher projektiven Ebenen. Arch. Math. 22 (1971): 214-220.
9. W. Feit. Finite simple groups, Proc. International Congress, Nice, 1970, vol. 1: 55-93.
10. M. Ganley. Polarities in translation planes. Geom. Dedicata 1 (1972): 103-116.
11. D. Goldschmidt. 2-fusion in finite groups. Ann. Math. 99 (1974): 70-117.
12. D. Gorenstein. Finite groups. Harper, New York (1968).
13. C. Hering. On subgroups with trivial normalizer intersection. J. Algebra 20 (1972): 622-629.
14. —— Eine Bemerkung über Streckungsgruppen. Arch. Math. 23 (1972): 348-350.
15. —— On shears of translation planes. Abh. Univ. Hamburg 37 (1972): 258-268.
16. —— On involutorial elations of projective planes. Math. Z. 132 (1973): 91-97.
17. A. Hoffer. On unitary collineation groups. J. Algebra 22 (1972): 211-218.
18. D.R. Hughes and F. Piper. Projective Planes. Springer New York (1973).
19. B. Huppert. Endliche Gruppen I. Springer, Berlin (1967).

20. C. Hering and W. Kantor. On the Lenz-Barlotti classification of projective planes. Arch. Math. 22 (1971): 221-224.
21. W. Kantor. On unitary polarities of finite projective planes. Canad. J. Math. 23 (1971): 1060-1077.
22. —— Those nasty Baer involutions. Proc. Internat. Conf. on Projective Planes at Washington State Univ. (1973): 145-155.
23. —— On the structure of collineation groups of finite projective planes (to appear).
24. E. Shult. On the fusion of an involution in its centralizer (to appear).
25. M. Seib. Unitäre Polaritäten endlichen projektiven Ebenen. Arch. Math. 21 (1970): 103-112.
26. H. Unkelbach. Eine Charakterisierung der endlichen Hughes-Ebenen. Goem. Dedicata 1 (1973): 148-159.

ADDITIONAL REFERENCES

R.C. Bose. On a representation of Hughes planes. Proc. Internat. Conf. on Projective Planes at Washington State Univ. (1973): 27-57.

T. Czewinski. Collineation groups containing no Baer involutions. Ibid. 71-75.

—— On collineation groups that fix a line of a finite projective plane (to appear).

W. Kantor. On homologies of finite projective planes. Israel J. Math. 16 (1973): 351-361.

M. O'Nan. Characterization of $U_3(q)$. J. Algebra 22 (1972): 254-296.

NOTES ADDED IN PROOF

1. In Theorem 1.5, hypothesis (1) should be replaced by the following condition. (1)' For each $(x,y) \in K \times K - \{(0,0)\}$, the matrix $\begin{bmatrix} x & y \\ u(x,y) & v(x,y) \end{bmatrix}$ is non-singular. (Condition (1)' is necessary for the existence of multiplicative inverses in F, but it is not obvious that (1)' follows from (1)-(4)). "Near-field functions", as defined on page 288, clearly satisfy (1)', and so do the

functions which occur in part (6) of the proof of Theorem 1.7. Thus the result of Theorem 1.7 is not affected.

2. When these notes were written, I was unaware that the special Hughes planes were already known to be self-dual. (L.A. Rosati: Sui piani di Hughes generalizzati e i loro derivati. Le Mathematiche 22, 289-302 (1967)). By using this result, the proof of (6) in Theorem 1.7 can be simplified. (See the remark after Theorem 1.7).

3. It was pointed out by Professor Lüneburg that there is a mistake in the published proof that the case $q = 19$ cannot occur in part (2) of the proof of Theorem 1.9. However, this can be rectified, and the result of Theorem 1.9 is correct.

4. A unified coordinate-free proof of Theorem 1.2 has now been given by Lüneburg ("Characterizations of generalized Hughes planes". To appear in Canad. J. Math.). He also shows there that in Theorem 1.1 "every perspectivity" can indeed be replaced by "every elation" when π is finite.